PRODUCTION AND UTILISATION OF SYNTHETIC FUELS— AN ENERGY ECONOMICS STUDY

PRODUCTION AND UTILISATION OF SYNTHETIC FUELS— AN ENERGY ECONOMICS STUDY

F. R. BENN, J. O. EDEWOR and C. A. McAULIFFE

Department of Chemistry,
University of Manchester Institute of Science and Technology,
Sackville Street, Manchester 1, UK

APPLIED SCIENCE PUBLISHERS LTD
LONDON

APPLIED SCIENCE PUBLISHERS LTD
RIPPLE ROAD, BARKING, ESSEX, ENGLAND

British Library Cataloguing in Publication Data

Benn, Frederick Roger
 Production and utilisation of synthetic fuels.
 1.   Synthetic fuel industry
 I.   Title   II.   Edewor, J O
 III.   McAuliffe, Charles Andrew
 662′.66   TP360

 ISBN   0-85334-940-1

WITH 88 TABLES AND 28 ILLUSTRATIONS
© APPLIED SCIENCE PUBLISHERS LTD 1981

Photosetting by Thomson Press (India) Ltd, New Delhi

Printed in Great Britain by Galliard (Printers) Ltd, Great Yarmouth

Preface

In recent years there has been a growing awareness of the importance of energy as shown in the large number of publications, many government sponsored, which attempt to project international energy requirements and availability into the 21st century. Added impetus for this resulted from the slump in industrial activity due to the dramatic increases in oil prices since 1973. The special vulnerability of western society to the shortfall has renewed interest in sources which are alternatives to the fossil fuels, coal, oil and gas, for example geothermal, tidal, nuclear and solar energy. However, at present only a limited effort is made, in view of their potential importance, to quantify them and so show to what extent they may be regarded as viable alternatives.

Energy analysts have shown that energy requirements of projects may be obtained in a related manner to their economic costs, and Chapman[3] emphasises the role of energy in the costing of these processes. An extension of this approach is of course process selection based on energy costs which could be important from the point of view of national energy conservation.

This work is concerned with the application of energy economics to the manufacture of synthetic liquid and gaseous fuels from coal, oil shale, tar sands and municipal and industrial wastes. The basic approach to process costing is to relate all costs incurred to energy, labour and capital—a three-factor model approach.

In this study we have been much influenced, and enlightened, by Dr David Merrick (National Coal Board Research, Cheltenham) and Dr David Hemming (Open University). For their interest and unselfish help we are deeply grateful.

F. R. BENN
J. O. EDEWOR
C. A. MCAULIFFE

Contents

APPENDICES

Units and Conversion Factors

To convert	To	Multiply by
joules	horse-power-hours	3.73×10^{-7}
joules	kilowatt-hours (kWh)	2.78×10^{-7}
joules	Btu	0.949×10^{-3}
joules	daily energy output from the sun	5.8×10^{-32}
joules	daily energy from the sun incident on the earth	6.7×10^{-23}
tonnes of coal	Btu(th)	27.8×10^{6}
short tons of coal	Btu(th)	25.3×10^{6}
tonnes of coal	therms	256.4
tonnes of coal	joules(th)	25.6×10^{9}
kWh(e)	megajoules(e)	3.6
kWh(e)	megajoules(th)	11.8
calories per gram	megajoules(th) per tonne	4.252
Btu per pound	calories per gram	1.80
scf of natural gas	Btu(th)	10^{3}
tonnes of oil	tonnes of coal	1.67
litres	cubic feet	0.0353
cubic metres	cubic feet	35.25
acres of land	square metres of land	4053
hectares of land	square metres of land	10^{4}
Btu(e)	Btu(th)	3.30
units of electricity	scf of natural gas	11.24

Introduction

The term energy economics here means the quantification of energy demands in the various processes yielding synthetic fuels and the use of these quantities to estimate costs of such processes. It is tempting for an economist to regard energy as a factor of production similar to a material or a piece of capital equipment. But in fact energy has a special role in the economy:[1]

(i) Energy is an essential input to every production, transport and communication process.
(ii) Energy is non-substitutable (in the sense that the best that can be achieved is the thermodynamic minimum energy input to a process, for which there is no alternative).
(iii) Ultimately the physical limits to man's activities on Earth will arise as a result of an energy constraint.

Most important of the roles of energy in the economy is the non-substitutability of energy and its necessity in every production process. Any other input to the production process can be substituted or reduced to an arbitrarily low level. For example, one material can be substituted for another; water can be obtained from different oceans; and the labour or capital inputs to a process can be reduced to very low levels. These manoeuvres, of course, have their respective financial penalties but they are technologically feasible. However, it is impossible to produce aluminium from alumina or ammonia from nitrogen and water without supplying at least the thermodynamically minimum amount of energy. Energy is thus a special input to production.

1

Having established the usefulness of energy as a tool of production, the next question is 'how useful is energy to the concept of costs in production?'. To answer this, one has to compare economic and energy analyses as techniques for product and process evaluation. The question of determining prices of commodities by economic analysis presumes that there is a 'perfect market'. But this is not always so and it has been recognised that there may be imperfections in the structure of the market, that there may be social costs excluded from prices and that external factors such as pollution or waste disposal costs may not be included. Also, other uncertainties about the future may occur because of this tendency to consider sub-systems of the economy. However, energy analysis attempts to eliminate this by considering the entire system, right from the resource stocks to some final product, including not only the final production industry but also those industries which supply it with raw materials, fuels, equipment, and so on. Thus, provided all input indices are considered with relevant energy demands being made at various stages, a more realistic cost analysis can be made and prices standardised according to energy involvement. It may be argued that various processes have varied levels of energy requirements as a function of production, but when consideration of the cost of any process in an industry can be expressed in terms of labour, energy and capital, then the above argument is valid. For example, cost C per unit production can be expressed as

$$C = X_f P_f + X_c P_c + X_l P_l \qquad (1)$$

where X_f = quantity of fuel (energy) (tonnes of oil equivalent per tonne of product);

P_f = price of fuel (energy) (£ per tonne oil equivalent);

X_c = quantity of capital (£ per tonne of product);

P_c = price of capital (£ per £ borrowed);

X_l = quantity of labour (man-hours per tonne of product);

P_l = price of labour (£ per man-hour worked).

Even the labour factor contains physical energy involvements, though this is relatively low compared to other energy inputs: as a convention we have given labour zero energy value.

It is being increasingly realised that energy quantification is not just a mere academic exercise but leads to constructive establishment of prices. Indeed, organisations engaging in input energy analyses find that financial savings can be achieved through savings in energy utilisation. From empirical data, based on experience, some workers[2] have come to realise

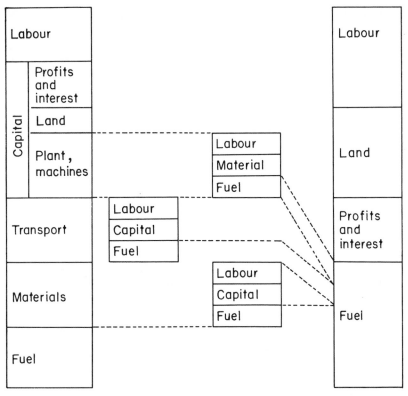

Fig. 1. The division of factor inputs into payments to personal incomes and payments for fuels. (Source: ref. 3.)

that cost per unit product is highly dependent on production energy intensiveness, i.e. the more the process is dependent on energy utilisation the more costly is the finished product. In conventional economic analysis some of these energy inputs are classified as 'overheads'.

Figure 1 shows all the factor inputs for the production of a unit of product (left block). In economic analysis these factors can be expressed as a set of financial costs, i.e. items that can be purchased and expressed in terms of money. Suppose the cost of producing a unit is expressed as C, and these items can be purchased as X_i units at prices P_i; then economic analysis expresses the cost of unit output as

$$C = \sum_i X_i P_i \equiv \sum_j P_j \tag{2}$$

But in energy analysis this factor breakdown is not extended to the payments on fuel energy inputs. Thus, energy analysis expresses costs of production (right block) as

$$C = \bar{X}_f \bar{P}_f + \sum_{j \neq f} P_j \qquad (2a)$$

where $\bar{X}_f$ = total quantity of fuel used
     = gross energy requirement of unit product (tonnes equivalent input);
  $\bar{P}_f$ = price of fuel used per unit product (£ per tonne of fuel input);
  P_j = personal incomes (excluding those ascribed to fuel inputs), in this case labour and capital.

One important difference that may arise in the methods of costing coal processes on the one hand and tar sands, oil shale and cellulosics on the other is that system boundary definitions will vary. In the case of coal processes the sub-system of conversion plant to finished product will be regarded as the considered system. In this case the energy requirements of mining, transportation, etc., will be regarded as already costed in the conventionally quoted coal prices of say £ (1978) 22·60 per tonne. Thus in these processes net energy recovery (NER) as product oil will not be meaningful. But $\bar{X}_f$ becomes $(1 + X_f)$ where X_f is defined as the ratio of NER of the process to the calorific value of product oil. Essentially X_f is equal to a fraction of the product oil (in energy equivalence) that is lost during the process of conversion. P_f becomes price of feedstock coal. This is dealt with in detail later.

With other processes the system boundary is assumed to include feedstock collection, through transportation, to withdrawal of product oil. In this case nothing enters the system but product oil is withdrawn. Moreover, the costs of collection are charged to both capital and labour, leaving the feedstock with no conventional market price. The energy losses that may occur, e.g. electricity utilised, chemicals, etc., are also assumed to be derivable from the product oil. Thus with NER/calorific value being X_f, the net energy recoverable from the process can be further defined as $(1 - X_f)$. The energy losses can only be costed at the on-site price of the product oil. To make this possible, an assumption can be made that a break-even situation should occur in these processes. These processes have $\bar{P}_f$ redefined as P_f, price of product oil, and $\bar{X}_f$ becoming X_f. Also, with the break-even assumption, cost of production, C, becomes price of product oil, P_f, i.e.

$$C = P_f$$

and eqn. (2a) becomes

$$P_f = X_f P_f + \sum_{j \neq f} P_j$$

and

$$P_f = \frac{\sum_{j \neq f} P_j}{(1 - X_f)} \tag{3}$$

i.e.

$$P_f = \frac{\text{non-fuel costs}}{\text{net energy yield of process}}$$

Equation (3) indicates that the economic viability of such processes depends on the value of X_f and hence P_f, which is a break-even price of product oil. Note that sociological cost elements such as taxation, grants, etc., are not considered. Thus, costs estimated here are essentially intended to represent accurately obtained 'cost-guides' to manufacturers employing these processes.

SYNTHETIC FUELS AND ENERGY ANALYSIS

Synthetic fuels are liquid and gaseous hydrocarbons derived from sources such as coal, shale, tar sands and very heavy crudes that cannot be processed by conventional techniques. Earlier we mentioned that the main drive in the world today to find ways and means of getting synthetic fuels is the unavoidable shortage of the present deposits of fossil fuels such as oil and gas. Thus in dealing with the energy economics of synthetic fuel manufacture, the general procedure would be

   (i) to evaluate the quantity and location of the resources for synthetic fuels,
  (ii) to evaluate the status of the recovery and upgrading techniques of these fuels,
 (iii) most important, to quantify the energy involvement of these techniques, the properties of the synthetic fuel products and the use of energy factors in estimation of the relative costs of the fuels.

Factors (i) and (ii) will be briefly dealt with here since most energy commentators touch on them. However, (iii), the costs and energy analyses, will be the main theme of this work. The range of resources will cover coal, tar sands, shale oil, waste plastics, cellulose and a mention of waste lubricating oil recycling is made. Part I of this book will treat the energy requirements of conversion to oil and gases of coal, oil shale, tar sands and general wastes while Part II will deal with the economics of these conversion processes.

First, we will survey the need to obtain oil from other resources. The efficiency of utilisation of oil and natural gas has made them surpass coal and other possible sources of fossil fuels as the primary fuels. However, the

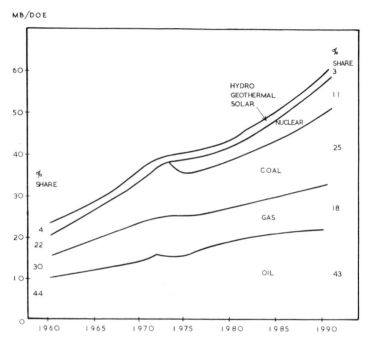

Fig. 2. US energy supply. Ordinate: million barrels per day oil equivalent. (Source: ref. 4.)

Growth rates (% per year)

	1960–73	1973–77	1977–90
Hydrothermal, goethermal and solar	4·6	2·4	0·4
Nuclear	40·8	31·7	13·0
Coal	1·8	2·5	5·3
Gas	4·3	(4·0)	0·2
Oil	4·4	1·2	1·9
Overall	4·0	0·5	2·8

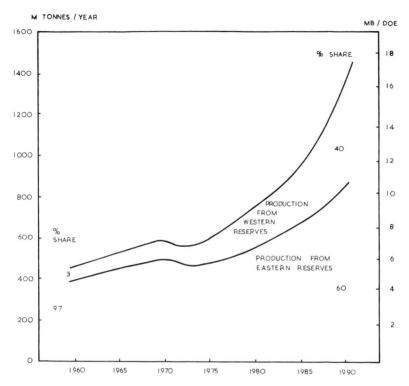

Fig. 3. *US coal supply. (92 million tonnes of coal per year = 1 million barrels per day oil equivalent.)*

Growth rates (% per year)

	1960–73	1973–77	1977–90
Western	12·8	20·7	11·7
Eastern	1·8	1·6	3·4
Overall	2·5	4·3	5·8

Growth rates are calculated on a tonnage basis and include exports and coal used in synthetics; they differ from those projections on energy supply, which are on a Btu basis. (Source: ref. 4.)

rates of consumption of these are exceeding the rates of production, as is evident in Figs. 2–5 and Tables 1–4.

Tables 1–3 contain estimates of British coal, gas and oil production until 2025 A.D. Current UK annual consumption of petroleum is about 82 million tonnes. The indications are that from 2025 A.D. oil and gas production will decrease while the production of coal may increase, suggesting that coal has a major role to play in energy contribution in

8   *Production and utilisation of synthetic fuels—An energy economics study*

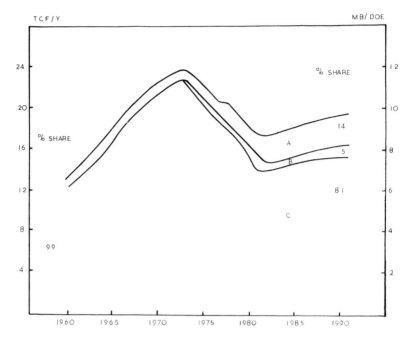

Fig. 4. US gas supply. (2 trillion cubic feet of gas per year = 1 million barrels per day oil equivalent.) (Source: ref. 4.)

Growth rates (% per year)

	1960–73	1973–77	1977–90
Imports (A)	15·7	(6·5)	9·7
Synthetics (B)	8·8	66·7	9·1
Domestic production (C)	4·0	(4.3)	(1·5)
Overall	4·3	(4·0)	(0·2)

Britain. Table 4 also gives the various projected costs of the contributions of coal and other sources to the generation of electricity. As of 2000 A.D. the price of oil will be approximately twice that of coal for generating electricity and that of fuel for thermal reactors will be highest of all. This, together with the foregoing argument of production, gives coal the leading role as energy carrier, over oil, by 2025 A.D. Thus it will be essential to see how coal can be processed to yield oil. The industrialised world runs on a hydrocarbon economy; oil is more convenient to use than coal. Synthetic gas can be burnt for heating homes, or converted to petrochemical products or reformed and made into methanol for multi-purpose uses. The justification of the need to convert coal into synthetic oil and gas is an object of this work.

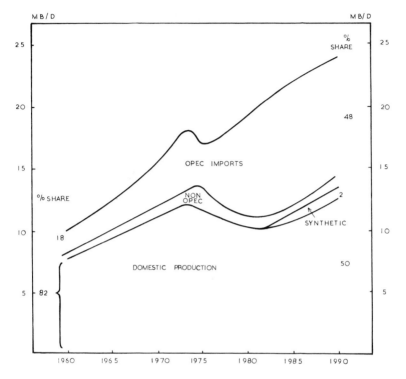

Fig. 5. US oil supply. Ordinate: million barrels per day oil equivalent. (Source: ref. 4.)

Growth rates (% per year)

	1960 – 73	1973 – 77	1977 – 90
Domestic production	2·5	(3·0)	1·2
Synthetics	–	–	30·5 (only 1985 – 90)
Imports	10·0	7·9	2·3
Overall	4·4	1·2	1·9

Figures 2–5 represent the US as a country with the largest coal deposits and still significant deposits of crude oil and natural gas. From these figures it appears that the 'synthetics' are expected to be of some significance in the US energy economy from 1985. In fact, the US leads all other countries in projects that have synthetic gases and oils as their final products. These include the biological conversion of cellulose to gaseous methane, the liquefaction of cellulose and the various gasification processes which exemplify US activities in the field of coal and fossil fuels conversion to

TABLE 1
Projected UK Coal Production (million tonnes)

Output from	1976	1990	2000	2010	2025
New mines	–	25	60	85	120
Existing mines (1976)	108	95	75	60	40
Open cast	12	15	15	15	10
Others[a]	2	–	–	–	–
Total	122	135	150	160	170

[a] Nearly 2 million tonnes of coal were recovered from dumps, under stagnant waters, etc.
(Source: ref. 5.)

TABLE 2
Projected UK Oil Production (million tonnes)

Output from	1976	1990	2000	2010	2025	Cumulative totals (1976–2025)
Proven and probable reserves	12	80	40	10	small	2 300
Additional reserves	–	30	60	60	20	1 600
Annual production	12	110	100	70	20	3 900

(Source: ref. 5.)

TABLE 3
Projected UK Natural Gas Availability (tcf)

Supplies from	1976	1990	2000	2010	2025	Cumulative totals (1976–2025)
UK continental shelf	3 800	5 000	4 000	3 000	500	65
Norwegian shelf	–	1 000	1 000	500	–	10
Other imports	100	small	500	500	small	5
Total supplies	3 900	6 000	5 500	4 000	500	80

(Source: ref. 5.)

TABLE 4
Cost of Electricity Generating Plants and their Fuels

Plant	Capital cost (£ per kW(e))[a]	Fuel cost
Nuclear	600	Uranium oxide at $40 per pound in 2000 rising to $75 per pound in 2025
Coal (PF)[b]	350	Coal at 13 pence per therm delivered in 2000 rising to 19 pence per therm in 2025
Gas turbine	200	Oil at 30 pence per therm delivered in 2000 rising to 50 pence per therm in 2025

[a] Capital costs include interest during construction, capitalised operation and maintenance charges, and also initial fuel charges (for nuclear) and connection to electrical grid.
[b] Pulverised fuel coal fired stations.
(Source: ref. 5.)

synthetic fuels. An interesting trend in Figs. 2–5 is that imports will be highly significant to the US from 1980; and, in order to alleviate this, the manufacture of synthetic fuels from coal will prove to be of great importance. Other consuming countries, and even producing countries such as Nigeria, should be thinking of how to replenish their depleting natural reserves of oil and gas. This then brings us to the various processes conventionally used to obtain synthetic fuels from coal, oil shale and tar sands.

Converting coal to synthetic fuels can be done by pyrolysis, liquefaction and gasification; obtaining oil from shale rocks can be done by underground room-and-pillar mining techniques, and highly efficient retorting processes; while with tar sands, those located close to the surface are surface-mined and the bitumen is extracted using hot water, steam and diluent. The deeper tar sands and the heavy oils must be recovered by *in situ* thermal stimulation, such as steam injection, to bring them to the surface. In both cases, the raw bitumens must be further upgraded to make a synthetic crude. Presently the only commercial synthetic crude operation on tar sands is the Great Canadian Oil Sands (GCOS) plant in Athabasca, Canada (see Section 3.1). The initial investment was about $253 million for a capacity of about 2·2 million tonnes per year, i.e. 45 000 barrels per day. Swabb[6] indicates that this plant has had many operating problems since it began production in 1967, but has made considerable improvement ever since, with a profit of $2·0 million being reported for the first half of 1974.

TABLE 5
In-place Reserves in Major Oil Sands Deposits in the World

Giant deposits (10^{12} barrels)		Large deposits (10^{9} barrels)		Medium deposits (10^{6} barrels)	
Location	*Size*	*Location*	*Size*	*Location*	*Size*
Orinoco (Venezuela)	0·7	Wabasca (Canada)	85	Selenizza (Albania)	371
Athabasca (Canada)	0·6	Peace River (Canada)	75	Guanoco (Venezuela)	62
Olenek (USSR)	0·6	Tar Sand Triangle (USA)	18	Asphalt Lake (Trinidad)	60
Cold Lake (Canada)	0·16	PR Spring (USA)	4	Santa Rosa (USA)	57
		Sunny Side (USA)	3	Sisquoc (USA)	50
		Bemolanga (Malagasy)	1.75	Asphalt (USA)	48
		Circle Cliffe (USA)	1·3	Tataros (Romania)	25
		Asphalt Ridge (USA)	1·2	Cheildag (USSR)	24
				Edna (USA)	16·6

(Source: ref. 7.)

TABLE 6
Comparison of Bitumen and Synthetic Crude

Properties	Bitumen	Syncrude
API gravity	8–9	35
Boiling range (°C (°F))	205–594 (400–1100)	27–482 (80–900)
Sulphur (weight %)	4·5–5·0	0·2
Nitrogen (weight %)	0·5–1·0	0·1
Vanadium (ppm)	150	Nil
Colour	Black	Straw
Ash (weight %)	1·0	Nil

(Source: ref. 8.)

TABLE 7
Synthetic Crude Production Rates

Year	Production (10^3 barrels per day)
1969	27·3
1970	32·7
1971	42·2
1972	51·0
1973	50·0
1974	45·7
1975	43·0

(Source: ref. 8.)

The plant's present capacity is about 3·0 million tonnes per year (1974 figures).

Tables 5–7 give some data on tar sand reserves, properties of syncrude obtained from tar sand deposits and rates of production of syncrude from tar sands over the years. The US is also investigating the prospects of the Utah sand deposits and it is reported that ERDA's efforts at development of the necessary *in situ* technology are progressing with reasonably encouraging results obtained in the first field experiments.[9] Scattered smaller deposits are also to be found in other countries (in particular the third world). This makes tar sands a useful energy source in the near future.

The US has the most commercial deposits of oil shale with the largest deposits being found in the states of Colorado, Utah and Wyoming. The estimated resource in place here is about 260×10^9 tonnes (1800 billion barrels) of oil. Of these, only about 7% (129×10^9 barrels) lie in deposits at least 9 m (30 ft) thick and average at least 125 litres of oil per tonne of shale (30 gallons per tonne). About 8×10^9 tonnes (54 billion barrels) of shale oil in this area can be recovered by the underground room-and-pillar mining technique and retorting processes. The details of such processes are discussed later but suffice it to mention here that two types of retorts are used—gas combustion and indirect heating. Some retort plants that are already in use that are worthy of mention are: two large scale gas combustion retort pilot plants—one of 330 tonnes per day, operated by the US Bureau of Mines between 1964 and 1966, and the other an 1100 tonnes per day retort, operated by the Union Oil Company between 1957 and 1958; in 1974 Petrosix demonstrated their hot recycle gas retort in a 2000 tonnes per day plant in Brazil. The Tosco II retort was also demonstrated in 1974 and is the basis on which the first US commercial shale oil plant is being planned. Other new projects that are either at developmental stages or are near demonstration stages are the 320–450 tonnes per day project organised by Paraho, a 1400 tonnes per day Union–SGR retort and a 22 tonnes per day plant planned by the Institute of Gas Technology of Illinois in which shale is retorted in a moderate pressure hydrogen atmosphere.

Cellulose and wastes (refuse) in general will also be considered, and the profitability, or otherwise, of treating municipal refuse and specially grown crops established. Most cities in the world that report refuse analyses seem to give figures which centre around 70% combustible materials in the refuse. Waste plastics are rising gradually in the refuse as packaging materials and by 1980 the proportion is estimated to be around 6·5%.[10] There are also materials such as tyres and lubricating oils from motor industries, and these will be quantified under the waste heading.

Of all the processes considered in Part I, particular attention will be paid to coal and cellulose, and of the coal processes currently being explored, much emphasis will be laid on the following: the Fischer–Tropsch gasification–synthesis processes of converting coal to methanol and motor spirit; the H-coal process devised by Hydrocarbon Research Incorporated (USA); the extractive hydrogenation or solvent refining processes, typical of the Consol Synthetic Fuels (CSF) process developed during the 1960s by the Consolidation Coal Company; and the pyrolysis or carbonisation processes, typical of the Char–Oil Energy Development (COED) process developed by the FMC Corporation (USA). We shall also look at two processes (study cases) being investigated by the National Coal Board Research Establishment, Cheltenham, for processing coal to synthetic liquids. These are the supercritical extraction process, which employs toluene as the carrier solvent, and the hydrogenative extraction of coal with anthracene oils. For cellulose-based processes, refuse as a feedstock will be considered first. Then the evaluation of specially grown crops will be made. Hemming[11-13] has carried out extensive calculations of energy requirements of processes converting coal, tar sands and oil shale to synthetic oil and gas. These calculations will be presented in summary form in the case of oil shale and tar sands, and more elaborately in the case of coal processes. Analyses will be made on cellulose materials and refuse in general. All resulting data will then be used to estimate costs for the various processes. Subsequently, comparative discussion will be carried out on coal and cellulose processes.

PART I

**Synthetic Fuels Production from
Other Sources of Energy—
Energy Requirements**

CHAPTER 1

Coal Conversion Processes

1.1 PREAMBLE

In general, the processes for converting coal to oil require an increase in the hydrogen–carbon ratio of the system. Thus, by addition of hydrogen atoms to coal molecules, a typical ratio of $0.8 : 1.0$ in coal is changed to about $1.75 : 1.0$ found in oils. Moreover, in the process of adding hydrogen to coal, some undesired components of coal, e.g. ash, moisture, oxygen, nitrogen and sulphur, may be eliminated, the last two as ammonia and hydrogen sulphide, respectively. The usual source of hydrogen is water (steam) and all the processes that will be considered are assumed to manufacture the required hydrogen internally. The energy required for this comes from the combustion of some part of the coal or products of liquefaction, pyrolysis or gasification.

Four basic routes are considered here: gasification–synthesis (Fischer–Tropsch), liquefaction by direct hydrogenation or by solvent refining, and pyrolysis (carbonisation).

1.2 BASIC PROCESSES

1.2.1 Fischer – Tropsch processes
In these types of reaction the coal is first gasified in a high pressure Lurgi gasifier as follows*:

*For simplicity, coal is represented by C.

19

$$C + CO_2 \longrightarrow 2CO \qquad (4)$$
$$C + H_2O \longrightarrow CO + H_2 \qquad (4a)$$
$$CO + H_2O \longrightarrow CO_2 + H_2 \qquad (5)$$

Carbon dioxide is obtained by burning coal in oxygen-enriched air and the resulting heat is used to initiate and maintain the reactions (4) and (5).

$$C + O_2 \longrightarrow CO_2, \ \Delta H < 0 \qquad (6)$$

The gaseous streams from reactions (4), (5) and (6) are passed through purification stages where H_2S and NH_3 are removed. Then the resulting gases are passed over a catalyst yielding liquid products ranging from methanol to hydrocarbons of high molecular weights. In these reactions, catalyst selectivity is essential and various catalysts are used to yield different products. Figure 6 gives a diagrammatic representation of methanol and motor spirit manufacture from coal by the Fischer – Tropsch method. Equations (7) to (10) give molecular reactions encountered in the processes. Equation (7) is often an undesired reaction and the product methane is passed to a reformer where it is converted mainly to H_2 and CO.

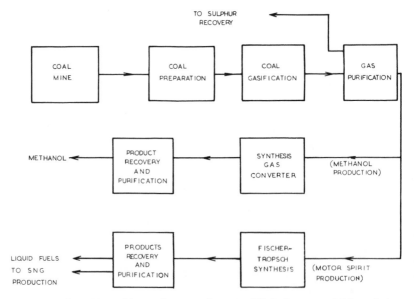

Fig. 6. *Fischer–Tropsch's synthesis paths—simplified diagram. (Adapted from ref. 14.)*

$$CO + 3H_2 \longrightarrow CH_4 + H_2O \tag{7}$$
$$nCO + 2nH_2 \longrightarrow C_nH_{2n} + nH_2O \tag{8}$$
$$nCO + (2n + 1)H_2 \longrightarrow C_nH_{2n+2} + nH_2O \tag{9}$$
$$CO_2 + H_2 \longrightarrow CO + H_2O \tag{10}$$

The Sasol process of South Africa is a typical example of the Fischer–Tropsch process yielding waxes, oils, motor fuel and a range of chemicals.

1.2.2 Hydrogenation processes

1.2.2.1 Direct hydrogenation
Depending on the pressure and residence times of the reacting coal and hydrogen, these processes yield products ranging from light hydrocarbons to heavy fuel oil. Usually these processes involve reaction of coal (slurried in recycle solvent) with hydrogen over a catalyst, and the reaction conditions are of the order 2000–4000 psi and temperatures of about 450 °C. Low pressures and short residence times yield heavy fuel while higher pressures and longer residence times favour the formation of lighter hydrocarbons. Typical yields are 2·5–3·5 barrels of oil and 2000–3000 scf (standard cubic feet) of gas per tonne of coal as given by Cochran.[15] Figure 7 illustrates the general scheme of a liquefaction process, of which direct hydrogenation is one type.

1.2.2.2 Extractive hydrogenation
This type of liquefaction reaction involves the partial or complete solubilisation of coal. Basically, two systems are being investigated commercially and their difference lies in the methods by which hydrogen is contacted with the coal. The solvent refined coal (SRC) system dissolves coal in an aromatic liquid in the presence of hydrogen gas under pressure of about 2500 psi. This results in the dissolution of most of the coal to form a slurry, which is filtered and the filtrate distilled to yield a product which is semi-solid when cool. This product usually has a low ash and sulphur content and is suitable as fuel for steam boilers, but can also be used as refinery feedstock. The second method is the hydrogen-donor process in which a solvent rich in hydrogen, e.g. tetralin or anthracene oil, is mixed with the coal and reacted under conditions of about 300 psi and 400 °C. The solvent extracts the volatile components of the coal, is cooled and filtered. The extracted components are recovered and the solvent regenerated and recycled. A typical yield is about 2–3 barrels of oil and 3500–4500 scf of gas

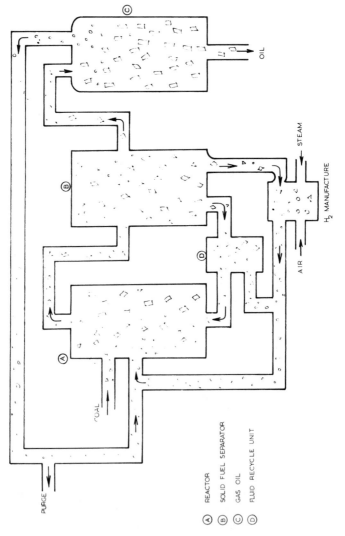

Fig. 7. Coal liquefaction.

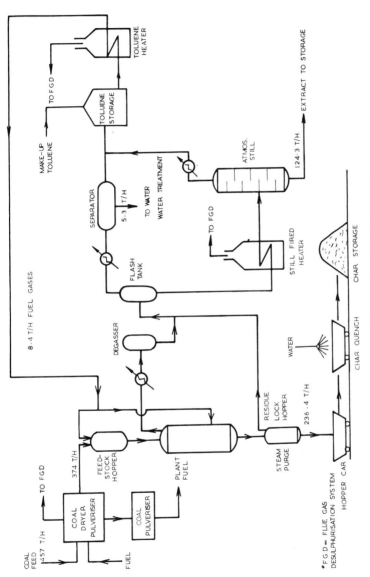

Fig. 8. Supercritical extraction of coal with toluene.

per tonne of coal. In both processes considerable processing is required to upgrade the initial products to syncrude.

At the National Coal Research Centre near Cheltenham, the liquefaction processes being investigated include the supercritical extraction of coal and the reaction of coal with anthracene oil as the extractive solvent. Figure 8 represents the supercritical extraction process, in which hydrogen is not required and toluene is the carrier solvent. The toluene is vaporised and its temperature raised to the supercritical level, and contacted with the coal. The volatile components pass into the toluene vapour medium and are carried along until the medium is cooled. At this stage, the particles of

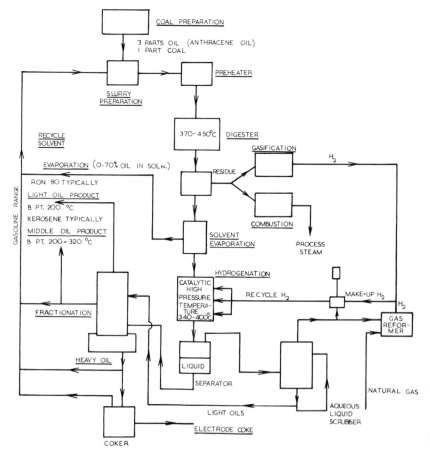

Fig. 9. Hydrogenative extraction of coal with anthracene oils.

coal extracted are dropped from the medium with the latter being reheated and recycled. In this process, the coal is usually heated to about 400 °C (750 °F) and any carrier medium with a supercritical temperature with a similar range can be used to extract the volatiles of coal. In a typical design estimate, about 457 tonnes per hour of coal gave about 237 tonnes of char, 124 tonnes of extract and 8·4 tonnes of fuel gases per hour.

In the process with anthracene oils, the reaction conditions are typically 400–480 °C, hydrogen pressures are about 200–340 bars and catalysts such as oxides of cobalt and molybdenum are used. Products are typically of the light oil range with boiling points less than 200 °C, the middle oil range with boiling points 200–320 °C and the heavy oil range with boiling points greater than 320 °C. Figure 9 shows this process schematically.

1.2.3 Pyrolysis processes
In this type of process the coal is rapidly heated in the absence of air; it decomposes, eliminating some volatile components containing carbon. In essence this carbonisation stage reduces the carbon–hydrogen ratio by the application of heat, producing char, gas and oil fractions. The char is removed and the gas–oil fractions are recovered as the main products. The oil is filtered to remove solids and then further hydrotreated to produce syncrude. The hydrogen used can be obtained by the reforming of the char produced, and the heat required for the process is obtained by burning either some coal input or some char products. The best known of the pyrolysis processes is the COED process. Typically, oil production is low—about 1 barrel of syncrude per tonne of coal. Char usually accounts for about 50–60% of the total product and can be a potential source of pipeline town gas (by an accompanying gasification process).

1.3 CONVENTIONS ADOPTED IN THIS WORK

In deriving the energy requirements of the various processes described above, the terminology that will be adopted follows the conventions suggested by the International Federation of Institutes of Advanced Studies (IFIAS) in which the following definitions are agreed.

(i) The process energy requirement (PER) is the sum of the fuel energy supplied to drive all the process stages within the considered system.
(ii) The gross energy requirement (GER) is the PER plus the gross heat of combustion of inputs that have alternative uses as fuels.

(iii) The net energy requirement (NER) is the GER less the gross heats of combustion of the products of the process.
(iv) Units of measurements are mixed but are generally SI units. Energy is expressed in joules mainly, although Btu will be used in some cases. Weights are expressed in tonnes and flow rates will be mainly per standard day.
(v) In all the processes considered, the boundary of the system will contain processes from entry of coal to the exit of syncrude for purification elsewhere. Also, in most of the processes, the electricity used, the gases burnt, the hydrogen used and other derived products/reactants will be taken as generated within the boundary of the entire system.

1.4 ENERGY REQUIREMENTS OF COAL PROCESSES

Calculations for these are shown in Appendices A1–A6 and are based on either conceptual designs (as in the H-coal process) or on actual industrial runs (as in the Sasol process). Hemming's[11] calculations are models for the following results.

1.4.1 Energy requirements of the H-coal process

1.4.1.1 Case 1: Using eastern coal of Illinois
Calorific value of coal = 25 530 MJ(th) per tonne
Moisture content of coal = 10%
Energy inputs to stages of process (10^6 MJ(th) per standard day)
    (a) Calorific value of coal feed (36 182 tonnes) = 923·74
    (b) Plant and capital equipment   =   8·25
    (c) Catalyst and chemicals   =   6·28
    (d) Slurrying solvent   = 120·00
                               Total: 1058·27

Percentage of total plant energy requirement apportioned to synthetic crude oil production (Appendix A1)   = 94·9%
Production rate of syncrude   = 100 000 barrels per standard day
                                = 14 452 tonnes per standard day
Calorific value of syncrude   = 42 700 MJ(th) per tonne

Therefore

$$\text{GER of syncrude} = \frac{1}{14\ 452} \times 0\cdot949 \times 1058\cdot27 \times 10^6$$
$$= 69\ 492\ \text{MJ(th) per tonne}$$

and
$$\text{NER} = 69\ 492 - 42\ 700$$
$$= 26\ 792\ \text{MJ(th) per tonne}$$

1.4.1.2 Case 2: Using western coal of Wyoming River
Calorific value of coal $\quad = 18\ 100\ \text{MJ(th)}$ per tonne
Moisture content of coal $= 33\%$
Energy inputs to stages of process ($10^6\ \text{MJ(th)}$ per standard day)
    (a) Calorific value of coal feed (52 439 tonnes) $= 949\cdot14$
    (b) Plant and capital equipment $\qquad\qquad = \quad 8\cdot16$
    (c) Catalyst and chemicals $\qquad\qquad\qquad = \quad 7\cdot17$
                                     Total: $\quad$ 964·47

Percentage of total plant energy
requirement apportioned to syn-
thetic crude oil production
(Appendix A1) $\qquad\qquad\qquad = 96\%$
Production rate of syncrude $\qquad = 13\ 538$ tonnes per standard day
Calorific value of syncrude $\qquad = 42\ 700\ \text{MJ(th)}$ per tonne

Therefore

$$\text{GER of syncrude} = \frac{1}{13\ 538} \times 0\cdot96 \times 964\cdot47 \times 10^6$$
$$= 68\ 392\ \text{MJ(th) per tonne}$$

and
$$\text{NER} = 68\ 392 - 42\ 700$$
$$= 25\ 692\ \text{MJ(th) per tonne}$$

The main reason for the difference in net energy requirements for the two cases is that the different coals used affect the quantities of hydrogen required for the process. In turn, the amount of energy required for the hydrogen generation stage (Appendix A1) will also vary, as will that required for drying the coal and other internal stages.

1.4.2 Energy requirements of the extraction (CSF) process
The product from this process is motor spirit with a calorific value of 46 830 MJ(th) per tonne. The basic calculations are shown in

Appendix A2, and are based on a conceptual design presented in the summary report of 'Project Gasoline' by NTIS.[16] Again two cases using the eastern and western sub-bituminous coals of the US, are considered.

1.4.2.1 Case 1: Using US eastern coal
Calorific value of dry coal = 27 170 MJ(th) per tonne
Moisture content of coal = 14·4%
Energy inputs to stages of process (10^6 MJ(th) per standard day)
    (a) Calorific value of coal feed (25 723 tonnes) = 610·93
    (b) Plant and capital equipment   =   4·86
    (c) Catalyst and chemicals   =   6·70
                                  Total: 622·49

Percentage of total plant energy
apportioned to motor spirit production (Appendix A2)   = 96·8%
Rate of motor spirit production = 7667·3 tonnes per standard day
Calorific value of motor spirit   = 46 830 MJ(th) per tonne

Therefore

$$\text{GER of motor spirit} = \frac{1}{7667\cdot3} \times 0\cdot968 \times 622\cdot49 \times 10^6$$

$$= 78\ 590\ \text{MJ(th) per tonne}$$
and
$$\text{NER} = 78\ 590 - 46\ 830$$
$$= 31\ 760\ \text{MJ(th) per tonne}$$

1.4.2.2 Case 2: Using US western coal
Calorific value of dry coal = 27 360 MJ(th) per tonne
Moisture content of coal = 24%
Energy inputs to stages of process (10^6 MJ(th) per standard day)
    (a) Calorific value of coal feed (27 581 tonnes) = 608·57
    (b) Plant and capital equipment   =   5·24
    (c) Catalyst and chemicals   =   5·67
                                Total: 619·48

Percentage of total plant energy
apportioned to motor spirit production (Appendix A2)   = 97·4%
Rate of motor spirit production = 7607 tonnes per standard day
Calorific value of motor spirit   = 46 830 MJ(th) per tonne

Therefore

$$\text{GER of motor spirit} = \frac{1}{7607} \times 0\cdot974 \times 619\cdot48 \times 10^6$$

$$= 79\ 318\ \text{MJ(th) per tonne}$$

and
$$\text{NER} = 79\ 318 - 46\ 830$$

$$= 32\ 488\ \text{MJ(th) per tonne}$$

1.4.3 Energy requirements of the pyrolysis (COED) processes

The basic pyrolysis of coal yields about 60% by weight of low value char and about 20% by weight of oil. Because of the low oil yield a modification of the process has been proposed in which the char is further gasified to produce pipeline gas. Shearer[17] produced a conceptual design and economic study for such a plant, combining the COED pyrolysis process with a low pressure version of the Kellogg molten salt process to gasify char and produce crude oil, pipeline gas and small quantities of light hydrocarbons, phenols and sulphur. The relevant calculations are shown in Appendix A3. In the second case, the straight pyrolysis process only is considered and the char is taken to storage.

1.4.3.1 Case 1: Pyrolysis with char gasification

Calorific value of coal   = 26 040 MJ(th) per tonne
Moisture content of coal = 10%
Energy inputs to stages of process (10^6 MJ(th) per standard day)
  (a) Calorific value of coal feed (28 455 tonnes) = 740·97
  (b) Plant and capital equipment   =   7·81
  (c) Catalyst and chemicals   = 11·66
                             Total: 760·44

Percentage of total plant energy
apportioned to syncrude production (Appendix A3)   =   39·74%
Rate of syncrude production   = 3896·4 tonnes per standard day
Calorific value of syncrude   = 42 800 MJ(th) per tonne

Therefore

$$\text{GER of syncrude} = \frac{1}{3896\cdot4} \times 0\cdot3974 \times 760\cdot44 \times 10^6$$

$$= 77\ 558\ \text{MJ(th) per tonne}$$

and
$$\text{NER} = 77\ 558 - 42\ 800$$

$$= 34\ 758\ \text{MJ(th) per tonne}$$

1.4.3.2 Case 2: Pyrolysis without char gasification
   Calculations are based on a study by Eddinger[18] and are shown in Appendix A3.
   Calorific value of coal   = 31 400 MJ(th) per tonne
   Moisture content of coal = 6%
   Energy inputs to stages of process (10^6 MJ(th) per standard day)
     (a) Calorific value of coal feed (10 039 tonnes) = 297·38
     (b) Plant and capital equipment   =   0·64
                                           Total: 298·02
   Percentage of total plant energy
   apportioned to syncrude produc-
   tion (Appendix A3)   =   42·7%
   Rate of syncrude production   = 2028·6 tonnes per standard day
   Calorific value of syncrude   = 42 800 MJ(th) per tonne

Therefore

$$\text{GER of syncrude} = \frac{1}{2028\cdot6} \times 0\cdot427 \times 298\cdot02 \times 10^6$$

$$= 62\ 731 \text{ MJ(th) per tonne}$$

and $$\text{NER} = 62\ 731 - 42\ 800$$

$$= 19\ 931 \text{ MJ(th) per tonne}$$

1.4.4 Energy requirements of Fischer – Tropsch processes
Calculations on these processes are based on the work of Chan[19] and this work is divided into two parts here. One deals with methanol as product while the other deals with motor spirit as product, which is typical of the Sasol process in South Africa. As in other processes most of the auxiliaries are assumed to be generated on-site, e.g. electricity, steam, plant fuel, etc. Data are shown in detail in Appendix A4.

1.4.4.1 Case 1: Motor spirit as product
   Calorific value of coal = 20 600 MJ(th) per tonne
   Energy inputs to stages of process (10^6 MJ(th) per standard day)
     (a) Calorific value of coal feed (31 135 tonnes) = 641·00
     (b) Plant and capital equipment   =   7·25
     (c) Catalyst and chemicals   =   3·92
                                     Total: 652·17
   Percentage of total plant energy
   apportioned to motor spirit pro-
   duction (Appendix A4)   =   46·91%

Rate of motor spirit production $= 2862 \cdot 0$ tonnes per standard day
Calorific value of motor spirit $= 47\,000$ MJ(th) per tonne

Therefore

$$\text{GER of motor spirit} = \frac{1}{2862 \cdot 0} \times 0 \cdot 4691 \times 652 \cdot 17 \times 10^6$$

$$= 106\,891 \text{ MJ(th) per tonne}$$

.and
$$\text{NER} = 106\,891 - 47\,000$$
$$= 59\,891 \text{ MJ(th) per tonne}$$

1.4.4.2 Case 2: Methanol as product

Calorific value of coal $= 20\,600$ MJ(th) per tonne
Energy inputs to stages of process (10^6 MJ(th) per standard day)
  (a) Calorific value of coal feed (28 904 tonnes) $= 594 \cdot 68$
  (b) Plant and capital equipment $\quad\quad\quad\quad = \quad 6 \cdot 42$
  (c) Catalyst and chemicals $\quad\quad\quad\quad\quad\quad = \quad 2 \cdot 94$
$$\text{Total:} \quad 604 \cdot 04$$

Percentage of total plant energy
apportioned to methanol produc-
tion (Appendix A4) $\quad\quad\quad\quad\quad = 69 \cdot 36\%$
Rate of methanol production $\quad\quad = 10\,121$ tonnes per standard day
Calorific value of methanol $\quad\quad = 23\,100$ MJ(th) per tonne

Therefore

$$\text{GER of methanol} = \frac{1}{10\,121} \times 0 \cdot 6936 \times 604 \cdot 04 \times 10^6$$

$$= 41\,398 \text{ MJ(th) per tonne}$$

and
$$\text{NER} = 41\,398 - 23\,100$$
$$= 18\,298 \text{ MJ(th) per tonne}$$

1.4.5 Summary and discussion of data on basic coal processes

Table 8 gives a summary of the various NERs for the conventional processes of converting coal to liquid fuels. It is evident that the net energy requirement of a particular process depends mainly on the type of coal, the quantity used and the calorific value. The lower the calorific value, the larger the quantity of coal required for the process and hence the greater the amount of hydrogen required for reaction with the coal. Hydrogen production is energy intensive and thus will have an appreciable effect on the overall energy requirements of the process.

An important parameter to be deduced from the tabulated figures is the ratio of NER of the product to its calorific value. This ratio is called the X_f

TABLE 8
NER Values of Various Basic Coal Processes

Process	NER value (MJ(th) per tonne of oil)
H-coal (Case 1)	26 792
H-coal (Case 2)	25 692
COED (with char gasification)	34 758
COED (without char gasification)	19 931
CSF (Case 1)	31 760
CSF (Case 2)	32 488
Fischer–Tropsch (motor spirit)	59 891
Fischer–Tropsch (methanol)	18 298

TABLE 9
Ratio of NER of Products to their Calorific Values

Process	X_f value
H-coal (Case 1)	0·627
H-coal (Case 2)	0·602
COED (with char gasification)	0·812
COED (without char gasification)	0·188
CSF (Case 1)	0·678
CSF (Case 2)	0·694
Fischer–Tropsch (motor spirit)	1·274
Fischer–Tropsch (Methanol)	0·704

of the process:

$$X_f = \frac{\text{net energy requirement of product}}{\text{calorific value of the product}}$$

Table 9 lists values of X_f for the basic coal processes. As is seen, most of the processes have X_f less than 1·0. However, the Fischer–Tropsch process for producing motor spirit has $X_f = 1·27$, and the pyrolysis process followed by char gasification has $X_f = 0·81$. The significance of this factor is that the higher the value of X_f the more energy intensive the process and the more costly it will be to produce a unit of synthetic fuel from coal.

From eqn. (3), a manufacturing process based on priceless feedstocks will break-even where

$$P_f = \frac{\sum_{j \neq f} P_j}{(1 - X_f)} \qquad (1)$$

The higher X_f is, the smaller the denominator and the larger the value of P_f to break-even, resulting in higher process costs.

The X_f values of the extractive hydrogenation (CSF) and direct hydrogenation (H-coal) processes are lower than those of pyrolysis and Fischer–Tropsch processes. The COED process without char gasification can be ignored since char with a low calorific value (CV) is the major product. Therefore, if energy savings are to be made in coal conversion, the production and utilisation of hydrogen in liquefaction would seem to be the most likely area to offer the greatest rewards. The Fischer–Tropsch process involves the breakdown to very small molecules $(CO + H_2)$ of a polymer structure (coal) and then recombining these to give motor spirit and a wide range of chemicals, a process with a thermal efficiency lower than 45%. The COED process with char gasification is a very good candidate for pipeline gas production, e.g. town gas.

The thermal efficiency (TE) of a process is defined as

$$TE = \frac{\text{total calorific value of products}}{\text{total value of input energies}}$$

This is calculated for all the processes, initially considering the primary products only and then considering the wide range of products produced in the various processes, in Appendices A1–A4 and data are tabulated in Tables 10–12.

TABLE 10
Thermal Efficiencies of Various Processes

Process	Thermal efficiency (%)	
	Primary product	With respect to all products
H-coal (Case 1)	61·0	64·3
H-coal (Case 2)	60·0	62·5
COED (with char gasification)	22·0	55·2
COED (without char gasification)	29·3	82·6
CSF (Case 1)	57·5	59·4
CSF (Case 2)	57·5	59·4
Fischer–Tropsch (motor spirit)	20·6	41·0
Fischer–Tropsch (methanol)	38·7	55·8

TABLE 11
Products from the Sasol Process

Product	Production rate (tonnes per standard day)	Total energy output (production rate × heat value, 10^6 MJ)
Motor spirit	2862·00	134·50
Diesel oil	170·00	7·40
Waxy oil	129·20	5·51
LPG	175·20	8·64
Naphtha, tar oil	1768·10	98·23
Acetone	28·90	0·88
Methanol	3·74	0·08
Propanol	52·71	1·77
Methyl ethyl ketone	7·31	0·25
i-Butanol	5·96	0·22
n-Butanol	17·52	0·63
n-Pentanol	4·08	0·15
Phenol	149·07	0·55
Ammonia	313·55	6·99
Sulphur	150·08	1·39
		267·2

(Source: ref. 11.)

TABLE 12
Products from the Methanol (Fischer–Tropsch) Process

Product	Production rate (tonnes per standard day)	Total energy output (production rate × heat value, MJ per standard day)
Methanol	10 120·90	233·61
Tar, oil, naphtha	1 654·73	93·15
Higher alcohol and dimethyl ether	48·57	1·68
Phenol	142·39	0·53
Ammonia	299·49	6·68
Sulphur	123·52	1·14
		336·79

(Source: ref. 11.)

The energy requirements shown in Table 8 do not include the energy used in mining of the coal and also that needed for transportation. These have been separated for ease of treatment since a typical energy usage figure for mining could be obtained direct from the activities of the NCB in the collieries. Transport energy will be quite small, relative to the actual process energies, but this will be considered also.

1.4.6 Case study processes for converting coal to synthetic fuels

Most of the coals used in the foregoing analyses have been coals from American and South African coal deposits. It is instructive to consider now the local situation, and so two case studies will be treated using British coals. Case 1 will be the process based on a report by Davies *et al.*[20] In this study coal (CRC 702) is extracted using anthracene oils at 400–480 °C and 200–340 atm partial pressure of hydrogen. Case 2 deals with coal (Markham Main 803) supercritically extracted with toluene, a situation in which hydrogen is not required and conditions approximate to 315–400 °C (600–750 °F) and 100 atm. Typical overall thermal efficiencies are 75·7% with the supercritical extraction and 60% with the hydrogenative extraction process.

1.4.6.1 Case 1: Hydrogenative extraction project of the NCB

Davies[20] gives a concise description of the project which in essence involves

extraction of coal with $\longrightarrow$ hydrogenation of coal
coal-derived solvents   extracts

$\downarrow$

refining of hydrogenation
products

Appendix A5 deals with the details of the process and calculations of energy requirements. The cost estimate analysis is shown together with other processes in Part II.

Calorific value of coal   = 32 950 MJ(th) per tonne
Moisture content of coal = 16%
Energy inputs to stages of process (10^6 MJ(th) per standard day)
  (a) Calorific value of coal feed (3157 tonnes) = 104·40
  (b) Oxygen requirements   =   0·95
  (c) Plant and capital equipment   =   1·44
  (d) Catalyst and chemicals   =   6·40
                            Total: 113·20

Percentage of total plant energy
apportioned to main products
(Appendix A5) $= 76.73\%$
Rates of main products production are:
   Gasoline $= 356$ tonnes per standard day
                       ($CV = 48\ 671$ MJ(th) per tonne)
   Gas oil $= 739.8$ tonnes per standard day
                       ($CV = 44\ 091$ MJ(th) per tonne)
   Electrode coke $= 109.6$ tonnes per standard day
                       ($CV = 19\ 520$ MJ(th) per tonne)
   Weighted calorific
   value of products $= 43\ 210$ MJ(th) per tonne

Therefore

$$\text{GER of products} = \frac{1}{1205.6} \times 0.7673 \times 113.2 \times 10^6$$
$$= 72\ 046 \text{ MJ(th) per tonne}$$

and
$$\text{NER} = 72\ 046 - 43\ 210$$
$$= 28\ 836 \text{ MJ(th) per tonne}$$

1.4.6.2 Case 2: Supercritical extraction of coal
Maddocks and Gibson[21] give a concise description of the process, which
essentially involves

coal extracted with toluene $\longrightarrow$ coal extract plus char

Appendix A6 deals with the details of the process and calculations of
energy requirements, and the cost estimates are shown under the appropriate section.
   Calorific value of coal $= 34\ 131$ MJ(th) per tonne
   Energy inputs to stages of process (10^6 MJ(th) per standard day)
      (a) Calorific value of coal feed (10 968 tonnes) $= 374.00$
      (b) Electricity $=\ \ 13.90$
      (c) Make-up toluene $=\ \ \ 0.95$
      (d) Plant and capital equipment $=\ \ \ 2.40$
      (e) Chemicals $=\ \ \ 1.53$
                                      Total: 392.78
   Percentage of total plant energy
   apportioned to main products $= 36.83\%$

Rate of product (extract) production = 2983·2 tonnes per standard day
Calorific value of extract = 36 726 MJ(th) per tonne

Therefore

$$\text{GER of extract} = \frac{1}{2983 \cdot 2} \times 0 \cdot 3683 \times 392 \cdot 78 \times 10^6$$

$$= 48\ 492\ \text{MJ(th) per tonne}$$

and
$$\text{NER} = 48\ 492 - 36\ 726$$
$$= 11\ 766\ \text{MJ(th) per tonne}$$

1.4.6.3 Summary and discussion of case study processes

(A) *Hydrogenative extraction*

NER of process	= 28 836 MJ(th) per tonne
Calorific value of products	= 43 210 MJ(th) per tonne
X_f value of process	= 0·67
X_c, capital per tonne product	= £ 17·3 per tonne
Overall thermal efficiency	= 60·0%
Coal daily feed	= 3157 tonnes
Calorific value of coal	= 32 950 MJ(th) per tonne

(B) *Supercritical extraction*

NER of process	= 11 766 MJ(th) per tonne
Calorific value of extract	= 36 726 MJ(th) per tonne
X_f value of process	= 0·32
X_c, capital per tonne product	= £ 17·5 per tonne
Overall thermal efficiency	= 75·7%
Coal daily feed	= 10 968 tonnes
Calorific value of coal	= 34 131 MJ(th) per tonne
Thermal efficiency (with respect to extract only)	= 27·9%

The supercritical extraction process at present affords about one-third of the coal feed as liquid product, so the process is attractive only if the large proportion of residual char is processed to pipeline gas. This combined process is under investigation at the National Coal Board Research Centre. The hydrogenative extraction process has a 60% efficiency and X_f value of 0·67 which make it look economically and energetically viable.

CHAPTER 2

Synthetic Crude from Oil Shale Rocks

2.1 PREAMBLE

The isolation of synthetic crude from oil shale rocks involves mining and crushing the rock, then retorting it and finally pre-refining the raw shale oil. Usually this syncrude product is a refinery feedstock. Figure 10 shows the flow diagram of the oil shale complex.

The detailed energy analysis of extracting oil from shale rocks will not be discussed here; for more details reference is made to Hemming.[13,22,23] A summary of data and procedures of the energy – economic analyses will, however, be given here, and the three-factor model approach to cost estimates adopted (see Part II).

In considering the efficiency of the retorting stage of the process five cases can be considered:

>*Case 1* assumes poor retorting efficiency, and yields are typically 80%
>of the Fischer assay* with products from the retort amounting
>to 30 gallons per tonne grade of shale rock. The exit temperature
>of spent shale is assumed to be 180 °C.

* Fischer assay is a technique that uses a micro-retort to determine the grades of oil shale rocks. Samples of crushed shale are heated to produce oil (which is collected and its volume determined) and gas (which is not accounted for) and spent shale. Because the Fischer assay does not account for the total hydrocarbon content of an oil shale, yields of oil from a commercial retort may exceed the Fischer assay if the retorting efficiency is greater than that achieved in the micro-retort. Fischer assays greater than 100% have been recorded for the Tosco II retort process. Note that the collected oil products can be expressed in terms of total calorific value and a reference Fischer assay in calorific value for the micro-retort = 100%.

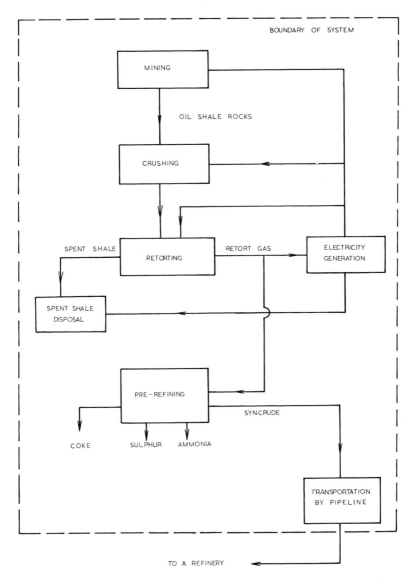

Fig. 10. Flow diagram of oil shale complex. (Source: ref. 13.)

Case 2 also assumes poor retorting efficiency with products having total calorific value equivalent to 90% of the Fischer assay for a retort of 30 gallons per tonne of shale rock; the exit temperature is again assumed to be 180 °C.

Case 3 is for fairly good retorting efficiency with products having total calorific value equivalent to 95% of the Fischer assay when retorting 30 gallons per tonne of shale rock.

Case 4 assumes good retorting efficiency with products having a total calorific value equivalent to 100% of the Fischer assay from the retort of 30 gallons per tonne grade of shale rock.

Case 5 assumes optimum retorting conditions. In this case it is assumed that heat or gas losses are negligible—with Fischer assays greater than 100%.

The purpose of the assumed efficiencies is to consider shale rock grades and retorting efficiencies expected in practical processes. Moreover, energy requirements will vary according to rock grade and retorting efficiencies. Energy requirements for the mining of shale grades are given in Table 13–15.

TABLE 13
Energy Requirements of Underground Mining with Explosives

Inputs	Energy requirements per tonne of shale		
	MJ(th)	*MJ(thi)*[a]	*MJ(e)*
Percussion drilling	5·23	–	2·08
Explosives	3·15	8·39	–
Scaling operations	0·29	0·70	–
Roof bolting	2·91	1·58	–
Mineyard and associated buildings	0·10	–	–
Mine ventilation	0·57	–	1·42
Water truck	0·01	1·93	–
Mine-road construction	1·50	–	–
Shovel loading	2·06	–	2·12
Truck haulage	4·65	18·80	–
Total PER[b] =	20·47 MJ(th) + 31·40 MJ(thi) + 5·62 MJ(e)		

[a] MJ(thi) = internally derived energy within the boundaries of system considered.
[b] Taking 1 MJ(thi) = 1·053 MJ(th), as defined by Swabb[6], total PER is given to be PER = 53·53 MJ(th) + 5·62 MJ(e).
(Source: ref. 13.)

TABLE 14
Energy Requirements for Underground Mining with Machinery

Inputs	Energy requirements per tonne of shale		
	MJ(th)	MJ(thi)	MJ(e)
Capital cost of MINER	3·46	–	–
Direct energy	–	–	9·37
Mining consumables	1·88	–	–
Machine maintenance	2·08	–	–
Scaling operations	0·29	0·70	–
Roof bolting	2·91	1·58	–
Mineyard and buildings	0·10		
Mine ventilation	0·57	–	1·42
Water truck	0·01	1·93	–
Mine-road construction	1·50	–	–
Shovel loading	2·06	–	2·12
Truck haulage	4·65	18·80	–
Total PER[a] =	19·51 MJ(th) + 23·01 MJ(thi) + 12·91 MJ(e)		

[a] Taking 1 MJ(thi) = 1·053 MJ(th), total PER = 43·74 MJ(th) + 12·9 MJ(e).
(Source: ref. 13.)

TABLE 15
Energy Requirements of Open-pit Mining

Inputs	Energy requirements per tonne of shale		
	MJ(th)	MJ(thi)	MJ(e)
Drilling	1·21	–	0·96
Explosives	1·94	5·16	–
Mine-road construction	3·00	–	–
Shovel loading	4·12	–	4·14
Truck loading	9·30	37·30	–
Total PER[a] =	19·57 MJ(th) + 42·46 MJ(thi) + 5·10 MJ(e)		

[a] Taking 1 MJ(thi) = 1·053 MJ(th), total PER = 64·28 MJ(th) + 5·10 MJ(e).
(Source: ref. 13.)

Unlike the case of coal processes, where relative transportation energy is almost negligible, the oil shale processes are transport intensive because of the loading/unloading of new shale rocks and the disposal of spent shales. The process is also labour intensive and it will be proper to consider the variation of energy requirements and costs with distance of retort plant from mining sites. Quite often retort plants are placed at a distance from the mines because of the nature of oil shale mines. In his calculations, Hemming[13] has assumed mine to retort distances of 15 km and retort to disposal distances of 15 km and 1 km.

To recap the process of obtaining oil from oil shales, the shale rocks are obtained either by underground mining with machinery, open-pit mining or mining by blasting of rocks with explosives. These rocks are then crushed near the mines and transported to the retort stills either by rail or trucks. Quite often trucks are used because of the variation of texture of shale mines (hardness or otherwise) with weather or climatic conditions of the area. The rocks are then retorted and the oil – gas recovered; spent shales (volumes of spent shales are usually greater than those of raw rocks) are then disposed of. The extracted oil is then pre-refined and passed on to a refinery as a feedstock.

There is no reported deposit of commercial importance of oil shales in the UK (the non-commercial deposits in Scotland are no longer worked) and summarised data were obtained from work based on the US Colorado oil shales.

2.2 ENERGY REQUIREMENTS OF THE STAGES OF AN OIL SHALE COMPLEX

2.2.1 Energy requirements of mining shale rocks

2.2.1.1 Underground mining with explosives
Hemming[13] presents details of underground mining with explosives. Room-and-pillar mining is generally accepted as a suitable underground method for oil shale. Table 13 gives the process energy requirements (PER) of this process. The major inputs in the mining operations are drilling, explosive materials, loading and haulage.

2.2.1.2 Underground mining with mining machinery
An alternative to blasting oil shale with explosives is the use of machinery underground. Hemming[13] has also estimated the PER of this process (Table 14).

2.2.1.3 Open-pit mining
Chapman[24] carried out an energy analysis of copper mining by open-pit

methods. His results provide a basis for estimating values for oil shale rocks (no study of open-pit mining of oil shale has been published).

2.2.2 Energy requirements of crushing stage

For each tonne of oil shale rock crushed the process energy requirement is approximately 14 MJ of thermal energy and 2·5 MJ of electrical energy.[13]

2.2.3 Energy requirements of retorting stage and disposal of spent shale

Hemming[13] estimated that the electrical requirement of retorting 1 tonne of shale would be about 18 MJ(e). Also, the retorting plant and associated facilities and utilities should have an amortised energy requirement of 18 MJ over a lifetime period of 20 years. Disposal energy is roughly 43 MJ of thermal energy and 3·4 MJ of electrical energy.

2.2.4 Energy requirements of pre-refining and sundry stages

The energy requirement of pre-refining raw shale oil is 4409 MJ(th) plus 270 MJ(e) per tonne of synthetic crude refined. One finds the equivalent of this per tonne of shale rock processed simply by considering the ratio of shale rock to syncrude. Pipelines and crude oil pumping to refinery account for about 324 MJ(th) per tonne of crude pumped.

In summary, therefore, the PER for production (mining to retorting) is 1188 MJ(th) + 280 MJ(e), while the PER for pre-refining and sundry processes is 4733 MJ(th) + 270 MJ(e).

TABLE 16
Net Energy Requirement of Syncrude from Oil Shale

Grade of oil shale (gallons per tonne of shale)	NER values for various retort cases (MJ(th) per tonne of syncrude)				
	Case 1	Case 2	Case 3	Case 4	Case 5
2·5	–	–	–	–	90 980
5·0	–	–	65 360	31 460	24 660
10·0	31 440	17 450	14 830	13 110	12 310
15·0	13 190	11 040	10 340	9 760	9 480
20·0	9 790	8 900	8 620	9 380	8 210
30·0	7 910	7 570	7 420	7 290	7 220
40·0	6 990	6 820	6 740	6 670	6 630
50·0	6 450	6 360	6 310	6 270	6 240
60·0	6 110	6 060	6 030	6 000	5 980

(Source: ref. 13.)

2.2.5 Net energy requirements of syncrude

These data are shown in Table 16. The five cases of Fischer assay were considered, together with the transport–disposal distances. The variations of NER of syncrude from oil shale with the grade of the shale rock are shown in Fig. 11.

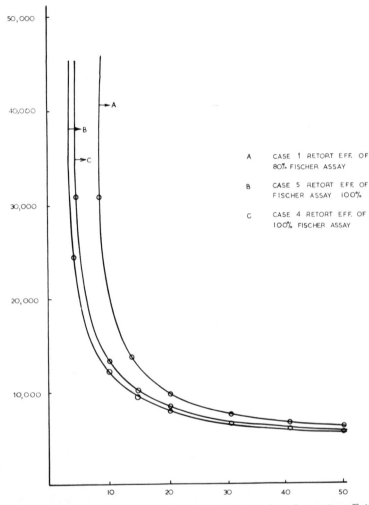

Fig. 11. Variation of NER of syncrude with oil shale grade and retorting efficiency. Ordinate: NER of syncrude (MJ(th) per tonne); abscissa: grade of oil shale (US gallons per short ton).

From Fig. 11 it is seen that production of syncrude is highly dependent on the grade of shale rock. Lower grades require an increase in the amount of shale which must be mined, crushed and retorted, and this is bound to affect the retorting heat required since more inert minerals have to be raised to the retorting temperature. Spent shale disposal is also a difficult problem. The net effect of this is that the NER of syncrude production rises rapidly as the yield falls below 10 gallons per tonne of shale rock. In this situation the extraction of oil from shale is uneconomical.

Of further interest is the actual value of the NER for various grades of oil shales. For grades of 15 gallons per tonne of shale and above a typical NER value is 7600 MJ(th) per tonne of syncrude; between 10 gallons per tonne and 15 gallons per tonne of shale, a typical NER value is 13 000 MJ(th) per tonne of syncrude. NER values for grades lower than 10 gallons per tonne of shale are very variable. For comparison, coal processes give NER values ranging from 8031 MJ(th) per tonne of syncrude in the pyrolysis process without char gasification, to 59 890 MJ(th) per tonne of motor spirit in the Fischer–Tropsch gasification–synthesis process. It is seen immediately that coal processes consume more energy than the shale process per tonne of product. However, the determining factor, as far as costs are concerned, should be the X_f value, i.e. the ratio of NER to calorific value of product. For coal processes, the hydrogenative extraction process (typical of the CSF processes) has X_f values of about 0·68 while those of shale processes range from 0·18 to 0·30. (The CSF process, however, produces gasoline

TABLE 17
X_f Values for Various Cases

Grade of oil shale (gallons per tonne of shale)	$X_f = \dfrac{NER}{\text{calorific value of syncrude}}$ (42 700 MJ(th) per tonne)				
	Case 1	Case 2	Case 3	Case 4	Case 5
2·5	–	–		–	2·131
5·0	–	–	1·531	0·737	0·578
10·0	0·740	0·409	0·347	0·307	0·288
15·0	0·310	0·259	0·242	0·229	0·222
20·0	0·230	0·208	0·202	0·220	0·192
30·0	0·180	0·177	0·174	0·171	0·169
40·0	0·164	0·160	0·158	0·156	0·155
50·0	0·151	0·149	0·148	0·147	0·146
60·0	0·143	0·142	0·141	0·141	0·140

while shale processes produce syncrude.) In both cases the mean calorific value of syncrude is taken as 42 700 MJ(th) per tonne. Table 17 gives the X_f values of various cases of retort efficiencies for varying grades of oil shale rocks.

CHAPTER 3

Synthetic Crude from Tar Sands

3.1 PREAMBLE

While the Great Canadian Oil Sands (GCOS) plant in Athabasca is so far the major commercial synthetic crude operation on tar sands, many other projects are being planned for the purpose of extracting oil from tar sands. These projects are based mainly on the technical lessons gained from the operational difficulties encountered with the GCOS plant. The latter, started in 1967, was not profitable until 1974 when it recorded a $ 2·0 million profit in the first half of the year. The initial investment was $ 253 million and anticipated capacity was 2·2 million tonnes per year (45 000 barrels per day). One of the new companies processing tar sands is Syncrude Canada Ltd, which is building a 6·3 million tonnes per year (125 000 barrels per day) mine and plant at a total investment of slightly over $ 1·0 billion (1974). With a work force of about 1400 men working at the Syncrude plant site, a start up at a capacity of 1·2 million tonnes per year (25 000 barrels per day) was planned for mid-1977. Also Shell Canada plans a 2·5 million tonnes per year plant to be on-stream by 1980. Others involved in this field are a group of five companies headed by Petrofina Canada Ltd and Home Oil/Alminex. By 1980 this operation as a whole should provide Canada with a total of 26 million tonnes output per year (515 000 barrels per day). This is a significant output of syncrude in terms of the national economy of Canada. Along with tar sand processing is the development of very heavy oils production in Canada. Thus bitumen, which is the major component of tar sands, can be seen as a future energy carrier for Canada.

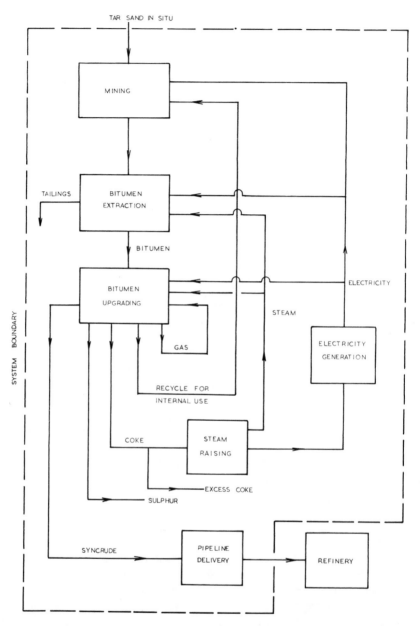

Fig. 12. Operations of the GCOS plant. (Source: ref. 12.)

However, empirical work has been done only on GCOS operations and the energy analyses and cost estimates will be based on the operations of this company.

3.2 OPERATIONS OF THE GCOS PLANT

Figure 12 gives a schematic flow of operations at GCOS. Mining activities on the tar sand mines are estimated at a level of 95 600 tonnes per calendar day. From these about 11 400 tonnes of bitumen are extracted daily, the rest being sand and other non-required substances. The mining technique is by the open-pit method, and a typical structural layout of a tar sand mine is thus: about 4 m of frozen muskeg (a semi-floating mass of root fibres), followed by approximately 16 m of overburden, mainly sand and other mineral deposits, and finally the tar sands, a mixture of mainly sand and bitumen. Thus to reach the tar sands, the 20 m of unwanted substances must be removed first. At the GCOS this is accomplished by the use of 150-tonne trucks.

The mining of tar sands is done by very large bucket-wheel excavators. The separation of bitumen from sand is accomplished by a hot water extraction process. The product bitumen is then upgraded by treatment with hydrogen in a number of delayed coking units. Typical final products of the upgrading are light gas, naphtha, kerosene, gas oil, coke and byproduct sulphur. A detailed process description is available.[12] One similarity between this operation and that of the treatment of oil shale rocks is that this process is also labour intensive. With a work force of over 1400 men at the plant site and with the movement of trucks to and from the mines, labour costs are expected to contribute a significant portion of total costs of producing syncrude from tar sands.

3.3 ENERGY REQUIREMENTS OF SYNCRUDE PROCESSING FROM TAR SANDS

Hemming[12] deals extensively with the calculations of energy requirements of this process based on GCOS operations. Summaries of the energy requirements will, however, be given here.

3.3.1 Energy requirements of mining processes
The mining process involves the stages of core sampling (to provide

TABLE 18

Stage	Energy requirement per tonne of tar sand mined
Core sampling	$2 \cdot 3 \times 10^{-3}$ MJ(th)
Muskeg removal	0·16 MJ(th) + 1·85 MJ(thi)
Overburden removal	0·48 MJ(th) + 5·66 MJ(thi)
Mining operation:	
(a) Bucket-wheel excavator	1·92 MJ(th) + 0·65 MJ(thi) + 1·50 MJ(e)
(b) Conveyor bridge	0·20 MJ(th) + 0·46 MJ(e)
(c) Conveyors	1·19 MJ(th) + 2·12 MJ(e)
Total[a]	3·95 MJ(th) + 8·16 MJ(thi) + 4·08 MJ(e)

[a] Assuming 1 MJ(thi) = 1·388 MJ(th), the energy requirement for the mining process = 15·30 MJ(th) + 4·08 MJ(e). (N.B. The figure 1·388 is given by Hemming[12] in the definition that 1 MJ(thi) = (1 + NER syn/42 400) MJ(th) where 42 400 is the calorific value of syncrude.)

information for the formulation of production plans), muskeg removal, main overburden removal and tar sands mining. The energy involvements in each of the stages are shown in Table 18.

3.3.2 Energy requirements of bitumen extraction
Detailed data and calculations are presented by Hemming[12] but the energy requirement associated with extraction of bitumen from tar sand is summarised in Table 19.

TABLE 19

Stage	Energy requirement per tonne of tar sand
Primary energy for steam and hot water	451·9 MJ(thi)
Electrical power for plant operations	11·2 MJ(e)
Chemicals	15·7 MJ(thi)
Amortised capital cost of plant	5·8 MJ(th)
Total[a]	5·8 MJ(th) + 467·6 MJ(thi) + 11·2 MJ(e)

[a] Taking 1 MJ(thi) = 1·388 MJ(th), the energy requirement for bitumen extraction = 655 MJ(th) + 11·2 MJ(e).

TABLE 20

Stage	Energy requirement per tonne of bitumen
Hydrogen production	2 264 MJ(thi)
Other processing plant fuel	3 982 MJ(thi)
Electrical requirement	138·7 MJ(e)
Catalysts and chemicals	66 MJ(th)
Capital equipment	232 MJ(th)
Total[a]	298 MJ(th) + 6 246 MJ(thi) + 138·7 MJ(e)

3.3.3 Energy requirements of bitumen processing and syncrude delivery

The energy requirement associated with the processing of bitumen at GCOS is summarised in Table 20.

About 80 MJ(thi) per tonne of syncrude is needed additionally to deliver syncrude to Edmonton.

3.3.4 Net energy requirement of syncrude produced at GCOS

From Sections 3.3.1–3.3.3 the total energy requirements can be summarised as follows:

(i) Mining $= 15·30$ MJ(th) + 4·08 MJ(e) per tonne of tar sand

(ii) Bitumen extraction $= 655$ MJ(th) + 11·2 MJ(e) per tonne of tar sand

(iii) Bitumen processing $= 8967·4$ MJ(th) + 138·7 MJ(e) per tonne of tar bitumen

(iv) Syncrude delivery $= 80$ MJ(thi) per tonne of syncrude

In calculating the GER and NER of syncrude, other assumptions were made by Hemming[12], viz:

(i) Overburden ratio to the tar sand was assumed to be 0·48 : 1·00.

(ii) Tar sands contained an average 13·26% by weight of bitumen.

(iii) Bitumen recovery was done at the plant with 90% efficiency.

(iv) All electrical requirements were derived internally by burning coke and raising steam at 30% efficiency, i.e. 1 MJ(e) $= 3·33$ MJ(thi) (as coke).

(v) 1 MJ(thi) $= (1 + $ NER syn$/42\,400)$ MJ(th).

From these assumptions and the energy requirements it can be calculated that

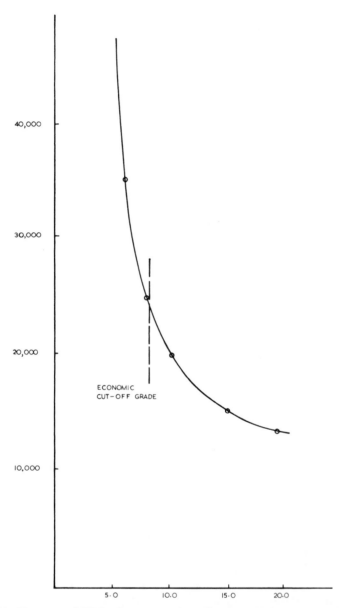

Fig. 13. Variation of NER of process with grade of tar sand. Ordinate: NER of process (MJ(th) per tonne of syncrude); abscissa: grade of tar sands (% bitumen). (Source: ref. 12.)

(i) $1 \text{ MJ(thi)} = 1 \cdot 388 \text{ MJ(th)}$

(ii) GER of syncrude $= 58\ 852 \text{ MJ(th)}$ per tonne and CV of syncrude is $42\ 400 \text{ MJ(th)}$ per tonne

(iii) NER of syncrude $= 58\ 852 - 42\ 400$
$$= 16\ 452 \text{ MJ(th)} \text{ per tonne}$$

(iv) X_f value of syncrude $= \dfrac{\text{NER}}{\text{CV}}$
$$= \dfrac{16\ 452}{42\ 400}$$
$$= 0 \cdot 388$$

3.3.5 Variation of NER with grades of tar sand

From Section 3.3.4 the NER for syncrude production is 16 452 MJ(th) per tonne of product for tar sands containing an average of 13·26% by weight of bitumen. This figure of 13·26 is typical of an average grade tar sand at the GCOS. But tar sands vary in grades and this variation is bound to affect the NER of the syncrude production process. Hemming[12] makes a concise study of these variations and Table 21 and Fig. 13 illustrate these. It will be

TABLE 21
Variation of Net Energy Requirements with Tar Sand Grade

Tar sand grade (weight % bitumen)	Weight of tar sand required to produce 1 tonne of bitumen (tonnes)	NER of syncrude (MJ per tonne)	X_f value of syncrude (NER/calorific value)
5·0	27·22	46 690	1·10
6·0	21·40	34 550	0·815
7·0	17·62	28 290	0·667
8·0	14·98	24 480	0·577
9·0	13·03	21 910	0·517
10·0	11·53	20 070	0·473
11·0	10·34	18 680	0·441
12·0	9·37	17 590	0·415
13·26	8·38	16 452	0·388
15·0	7·31	15 400	0·363
18·0	6·00	14 090	0·332
20·0	5·36	13 470	0·318

(Adapted from ref. 12.)

noted that an economic grade of tar sand to work with will be in the range of 7·50–8·50% by weight of bitumen. Below this range, the NER of the process tends to vary very sharply with slight decrease in tar sand grade. Table 21 also gives X_f values for various grades of tar sand. Figure 13 reveals that the exploration and production of syncrude from tar sands of grades below 7% by weight of bitumen may turn out to be quite uneconomical. At the cut-off grade of 8%, the cost of producing syncrude per tonne (Part II) is estimated to be as high as £ 100·00. This is equivalent to about $ 28·0 per barrel product. At the GCOS where the average grade is about 13·26% by weight the estimated cost of producing syncrude is about £ 60·00 per tonne.

CHAPTER 4

Conversion of Cellulose to Synthetic Oil: Liquefaction Process

4.1 PREAMBLE

The conversion of cellulosic materials to synthetic fuels, mainly syncrude, is a newer process than those concerning tar sands, coal and oil shale. Though much laboratory work has been done on cellulose conversion, and various results reported, the possible commercial viability of such processes is a recent idea. Essentially, cellulose is represented by the formula $C_6H_{10}O_5$, though the carbon–hydrogen ratio can vary. Carbohydrates and various sugar compounds have similar formulae and the actual stoichiometry of cellulose reactions is complex.

In converting cellulose to oils and gases it is thought that the cellulosic material is essentially hydrogenated. Essentially two reactions occur in the process. It is believed that an *in situ* water gas shift reaction occurs between the cellulose and water yielding carbon dioxide, carbon monoxide and hydrogen. Then the hydrogen reacts with excess cellulose to yield the final products:[25]

Shift reaction	$C + H_2O \longrightarrow H_2 + CO$	
Liquefaction	$CO + R—O \longrightarrow R— + CO_2$	
	$H_2 + R—O \longrightarrow R— + H_2O$	

i.e. the total process is

$$C + 2R—O \longrightarrow 2R— + CO_2$$

An attempt to substitute figures for a simple stoichiometric equation will give[25]

Shift	$5\,C_6H_{10}O_5 + 35\,H_2O$	$\longrightarrow\ 30\,CO_2 + 60\,H_2$
Liquefaction	$12\,C_6H_{10}O_5 + 60\,H_2$	$\longrightarrow\ 12\,C_6H_{10} + 60\,H_2O$
Total process	$17\,C_6H_{10}O_5$	$\longrightarrow\ 30\,CO_2 + 12\,C_6H_{10} + 25\,H_2O$

C_6H_{10} simply represents the loss of oxygen from the cellulose molecule but the products obtained will be mixtures of hydrocarbons in the gas, gasoline and gas oil ranges.

Reaction conditions, by experience,[25] tend to be quite severe. Temperatures range from about 352 to 454 °C for good yields, and pressures are in the range 30–100 atm. Under these conditions it is thought that hydrogen 'plucks' the oxygen atoms, usually existing as 3(—OH) and 2(—O) in the cellulose, leaving the hydrocarbon fragments to reunite to form the products. Further hydrogenation is required to make saturated hydrocarbons. Two case studies in the hydrogenation of cellulose to oil are described: the liquefaction of municipal 'refuse' and of specially grown crops. The details of the processes will not be discussed but relevant references will be given.

4.2 CONVERSION OF REFUSE TO OIL

4.2.1 Preamble of the Holden Town case study

A study of the Holden Town process of converting municipal refuse to oil, by cellulose liquefaction, was carried out by Kaufman and Weiss[25] in the US. Cellulose was one of the materials used for an integrated study by these workers, but other materials such as polystyrene were also used in preliminary studies. Powdered newspaper was taken as representative cellulosic material. It was found that with polystyrene degradation various monomers, dimers and trimers of the parent feedstock were obtained. Powdered newspaper yielded about 2·1 barrels of oil per tonne of feed. With these data a study was carried out on the municipal refuse of Holden Town in the US. With a weight composition of 75% of combustible materials in the refuse, and cellulosics accounting for about 60–65% of these, Kaufman and Weiss found that a net yield of about 2·1 barrels of oil was also obtainable per tonne of combustible material treated. It is probable that the plastics and rubbery materials simply decomposed to yield various oils and gases while the cellulosics engaged in the reactions described in Section 4.1.

According to Kaufman and Weiss,[25] material with a heating value in the range of wood (refuse), i.e. 21 300 MJ(th) per tonne can be used to make

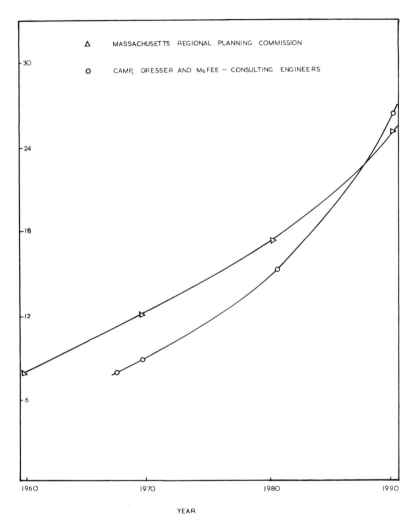

Fig. 14. Annual refuse production (tonnes × 10³). (Source: ref. 25.)

relatively pure oil having a calorific value of 43 000 MJ(th) per tonne. The town of Holden, where the study was carried out, has a refuse disposal system with annual amounts as shown in Fig. 14. Table 22 gives a weekly refuse distribution pattern as presented by Kaufman and Weiss. With their plant design it is estimated that a daily refuse throughput of 36 tonnes yields about 13 tonnes of oil daily. However, a scale-up procedure is

TABLE 22
Weekly Refuse Distribution as Percentage of Total

	Tonnes	Cubic yards	Vehicles
Monday	12·0	8·5	7·0
Tuesday	9·0	7·0	3·5
Wednesday	12.0	12·5	8·0
Thursday	7·0	7·0	3·5
Friday	10·0	10·0	11·0
Saturday	30·0	32·0	32·0
Sunday	20·0	23·0	35·0

available to give treatment of up to 2000 tonnes per day of combustible refuse.

The plant is divided into four main systems: the fed refuse preparation system; the hydrogen production and recovery system; the refuse liquefaction system; and the catalyst recovery and water treatment systems. Figure 15 gives a schematic of the plant and Fig. 16 shows the material flow pattern. The raw refuse is prepared by a mechanical air separator technique, as shown in Fig. 17, to obtain the combustible feedstock. This is ground by a hammer mill and stored in a silo until it can be slurried in oil. The slurried feed (refuse slurried in recycled product oil) is passed through a furnace for pre-heating and then into a reactor at about 452 °C and under 70 atm of H_2. The residence time of reactants in the reactor is usually 15·1 min for optimum oil yields. The catalyst employed in this process is nickel. The products are mainly gases in the range of C-1 to C-5, light hydrocarbon oils, and heavy oils. The gases and light oils are burnt to provide heat for the system leaving the heavy oils as the main product. The system can be self-sustaining in terms of energy/fuel requirements.

4.2.2 Energy requirements of a plant producing oil from municipal refuse

4.2.2.1 Input energy
The calculations of the energy inputs in the processing of oil from municipal refuse are shown in Appendix B1, and a summary of the requirements is given in Table 23. These are based on a 36 tonnes per day refuse capacity.

The gross energy requirement of the entire process is

$$943 \times 10^3 \text{ MJ(th) per day}$$

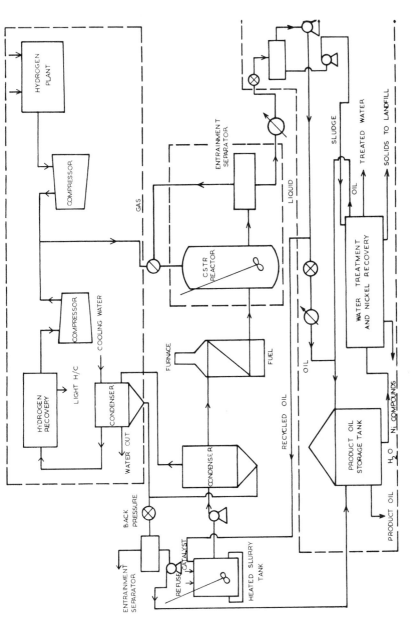

Fig. 15. Cellulose liquefaction plant. (Source: ref. 25.)

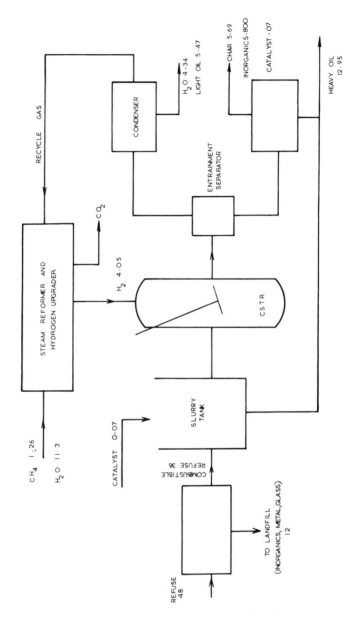

Fig. 16. Overall material balance (cellulose liquefaction). (All numbers given in tonnes per day.)

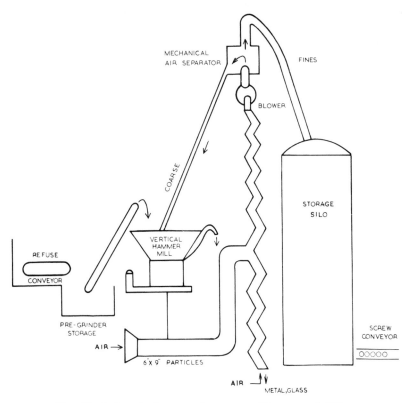

Fig. 17. Refuse feed preparation system. (Source: ref. 25).

TABLE 23

Component operations	Energy requirements (10^3 MJ(th) per day)
Electricity for grinding, pumping, etc.	50·94
Plant and capital equipment	15·16
Catalysts and chemicals	13·28
Hydrogen production	14·74
Boiler and furnace heating	82·07
Refuse (heat value)	766·80
Refuse collection[a]	—
Total	943·0

[a] Evaluated as transportation labour cost—see Part II.

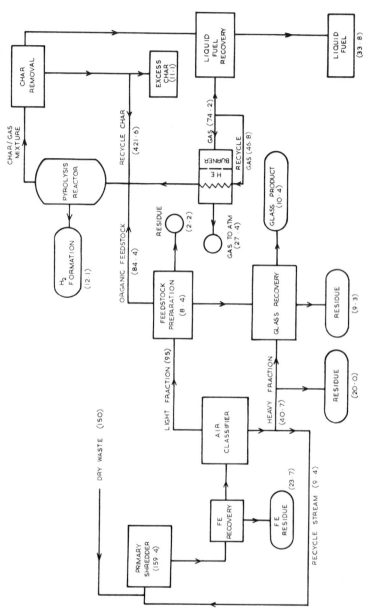

Fig. 18. Process outputs. (All numbers given in tonnes per day.)

4.2.2.2 *Process outputs and NER evaluation*

From Fig. 18, outputs of the process are estimated to be

(i) 12·95 tonnes per day of heavy oil ($CV = 43 \times 10^3$ MJ per tonne);
(ii) 5·69 tonnes per day of char ($CV = 19·5 \times 10^3$ MJ per tonne)
(iii) some water, inorganics and CO_2

In apportioning the energy requirements, the CO_2 and inorganics will be neglected. The CO_2 has no heat value and the inorganics are not considered here. The resulting products will thus be char and heavy oil. Therefore

% of total GER apportionable to oil production

$$= \frac{12·95 \times 43 \times 10^3}{(12·95 \times 43 \times 10^3) + (5·69 \times 19·52 \times 10^3)}$$

$$= 83\%$$

and

$$\text{GER due to product oil} = \frac{1}{12·95} \times 943 \times 10^3 \times 0·83$$

$$= 60·70 \times 10^3 \text{ MJ(th) per tonne}$$

$$\text{NER of product oil} = \text{GER} - \text{CV}$$

$$= (60·70 - 43·0) \times 10^3$$

$$= 17·7 \times 10^3 \text{ MJ(th) per tonne}$$

The X_f value fraction of oil product necessary as the required energy input for the process to take place is given by

$$X_f = \frac{\text{NER of oil}}{\text{heat value of oil}}$$

$$= \frac{17·7 \times 10^3}{43·0 \times 10^3}$$

$$= 0·41$$

Thus the net recoverable fraction of product oil $(1 - X_f)$

$$= 1 - 0·41$$

$$= 0·59$$

The thermal efficiency of the process with respect to the oil product only

$$= \frac{12·95 \times 43 \times 10^3}{943 \times 10^3}$$

$$= 59\%$$

TABLE 24

Plant refuse capacities (tonnes per day)	Plant capital costs ($ million)	Plant energy factor (X_f)
36	1·487	0·410
100	2·742	0·382
500	7·214	0·378
1000	10·914	0·375
2000	16·504	0·370

and with respect to char and oil

$$= \frac{668 \times 10^3}{943 \times 10^3}$$
$$= 71\%$$

In Appendix B2 the energy requirements of plants with capacities greater than 36 tonnes per day are calculated. Table 24 shows the summary of such plants with their various capital costs and energy factor values, X_f. It will be noted that there are slight differences in the X_f values. This is because a large fraction of total energy input to the plants comes from the refuse materials.

4.3 CONVERSION OF SPECIALLY GROWN CROPS TO OIL

4.3.1 The future of specially grown crops as energy carriers—an assessment

Kaufman and Weiss[25] have shown that the process for cellulose (refuse) liquefaction has potential for providing useful liquid and gaseous fuels. With a net fractional recovery of about 0·6 of intrinsic energy in the refuse at a very low cost (see Section 4.3.2.4) local refuse could be changed from the conventional status of 'rubbish' to a usable material. If this is acceptable universally as plausible then it is pertinent for one to look at other sources of cellulose material. Specially grown plants (grasses, trees, etc.) come to mind immediately. Solar energy is continuously turned into stored chemical energy in plants via the process of photosynthesis. In plants where the stored energy cannot be used as food the material (cellulose and related compounds) can be converted to oil and gases for use. This may sound rather optimistic but the fact is that cellulose

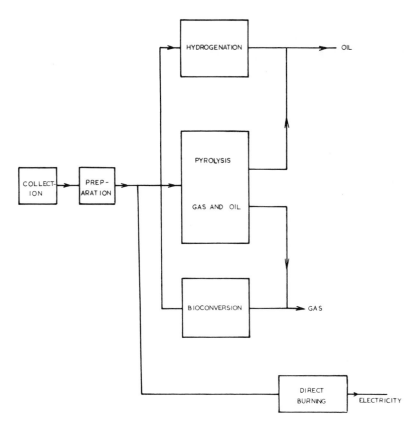

Fig. 19. Organic waste resource development. (Source: ref. 28.)

can comfortably compete with coal or oil shale as energy carriers for the future.

A report by the US Federal Energy Administration[28] presented a scheme (Fig. 19) by which organic wastes can be developed to energy forms. This essentially shows that organic wastes should be given consideration when future energy carriers are reviewed.

However, the determining factor for a future energy carrier is the abundance and hence the sort of continuity or otherwise of the source of energy. Municipal/industrial wastes are quite abundant and could form the basis of an allied energy technology. But local population wastes disposal patterns vary with time, level of social sophistication and other technological trends. Thus one has to look at the possibility of growing crops specially for conversion to oil. These crops could even be basic plants, in particular in areas where sunshine is abundant.

The United States is one of the countries currently investigating the potential of these continuously renewable energy sources as a means of combating the worsening energy situation. Sugar cane is one of the crops already providing energy as motor fuel. This fuel, gasohol, currently being used in the US is a blend of 10% alcohol from sugar cane and 90% conventional motor spirit. It has even been demonstrated that certain methanol–water mixtures have high octane ratings. Methanol can be obtained from bioconversion of cellulose to methane and conversion of the latter to alcohol. Other crops could be converted to various usable energy forms. Sitton and Gaddy[29] indicated the attractiveness of biological conversion of crops to methane with average annual returns on investments of about 19–27%. Thus one should regard grown crops as an energy source having great potential.

4.3.1.1 Some relevant data
The leaf canopy system of any plant serves primarily as a solar energy collecting surface, the effectiveness of which depends on the structure of the leaf collection surface and the intensity of the sun's energy at the surface of the leaf. Penner and Icerman[30] indicate that the mean solar power input to the Earth's surface is about 5.2×10^{21} Btu per year or about 5.5×10^{18} MJ per year. Of this about 0·023% is used for photosynthesis. This gives a mean total apparent solar energy recovery of about 1.27×10^{15} MJ per year. Sitton and Gaddy[29] give typical thermal conversion efficiencies of crops ranging from about 1·0% for corn and sugar cane to about 0·2% for a forest. Table 25 gives a summary of these conversion efficiencies, based on statistical data from US crops. Sitton and Gaddy also state that for any 14 daylight hours, during 1975, an amount of energy equal to about 84.4×10^{13} MJ was incident upon the surface of the US in the form of solar radiation. This is equivalent to an insolation rate of 3·2 MJ per minute-foot^2.

From the work of Alich and Inman,[31] Sitton and Gaddy constructed Table 26 of cropland areas and type of croplands in some states of the US. According to the latter, 10% of the cropland was idle in 1973, 87·0% of which was kept idle purposely for the US Federal Government. An estimate of this idle land was put at 19·2 million acres. Alich and Inman estimate that about 20 tonnes per acre-year of crops could be obtained on these lands, giving a mean total of about 380 million tonnes of biomass each year. At an average heating value of 15.4×10^3 MJ per tonne of

TABLE 25
Solar Conversion Efficiencies of Some Crops

Plant type	Fuel value (10^6 J per kilogram)	Dry yield (kg per m^2-year)	Estimated solar energy conversion (%)
Oak – pine forest	16·25	1·21	0·41
Southern pine	16·25	0·45 – 1·12	0·13 – 0·33
Hybrid poplar	13·06	0·90 – 1·80	0·24 – 0·47
Sycamore	13·46	0·36 – 2·51	0·09 – 0·61
Reed canary grass	15·09	1·42	0·29
Bermuda grass	13·06	1·80 – 2·47	0·42 – 0·58
Alfalfa	15·09	0·64	0·18
Corn	15·09	2·51 – 4·02	0·72 – 1·15
Sugar cane	15·09	4·49	1·11
Cattail swamp	15·09	2·51	0·88
Marine algae	15·09	2·02 – 2·63	0·63 – 0·74
Sewage pond	15·09	5·63	1·34

(Data provided by McCloud[26] and Kemp and Szego.[27])

TABLE 26
Distribution of Areas Classified as Croplands in the US

Type of cropland	Area (km^2)	Percent of total
Field crops	475 875	65·8
Other crops	76 001	10·5
Summer fallow	63 901	8·8
Hay land	29 623	4·1
Idle lands		
Conservation use	67 502	9·3
Temporary use	6 758	0·9
Open	3 440	0·5
Total free	77 700	10·8
Grand total	723 100	100·0

(From ref. 29. Data obtained from USDA—Statistical Report No. 461, 1969.)

biomass or crops, recoverable energy from such a programme could be put at $5·27 \times 10^{12}$ MJ per year. Anderson[32] showed also that about 400 million tonnes of agricultural wastes are available each year in the US. At a heating value of the same order it can be estimated that about $2·4 \times 10^{12}$ MJ per year are also recoverable from crop wastes in the US. With a potential as high as 800 million tonnes of grown crops and crop wastes supplementing municipal/industrial wastes, cellulose liquefaction provides an enormous potential to the US in terms of energy renewal in the form of oil and gas.

In the UK conditions are quite similar. Schomburgh[33] noted that in 1975 Britain was producing 16·5 million tonnes of domestic/commercial refuse per annum. This amounted to about 5·5 million tonnes of coal equivalent with an average heat value of $11·0 \times 10^3$ MJ per tonne. Also Bidwell and Mason[34] estimate that energy recoverable from Greater London Council's refuse, after materials (non-combustible) reclamation, is of the order of $8·34 \times 10^3$ MJ per tonne of refuse. In his description of photosynthesis as an energy converter, Cooper[35] noted that during 1973 the solar energy impinging on cultivable land in the UK was of the order of 450×10^{12} MJ. He noted that, though the emphasis of agricultural authorities will be on consumable crops such as wheat and vegetables, much crop wastes and residues occur in the process of planting and harvesting. He is of the opinion that these waste materials can be converted to usable energy. In 1973, for instance, $9·3 \times 10^6$ tonnes of dry matter were burnt as wastes. The heat equivalent of this is 64×10^9 MJ. He also estimates that another $27·5 \times 10^9$ MJ can be claimed from timber crop harvesting and a further 9×10^9 MJ per year from timber forests. With a total of about 100×10^9 MJ per year supplementing $60·5 \times 10^9$ MJ of wastes the process of cellulose liquefaction looks attractive to the UK. Tables 27 and 28 give the fractions of some cellulosic material outputs that are wasted, burnt or

TABLE 27
Comparative Energy Output and Utilisation of Crops
(MJ $\times 10^3$ per hectare)

Crops (country)	Total	Direct human food	Rest
Wheat (UK)	142	67	75
Potatoes (UK)	178	134	44
Timber (UK)	170[a]	–	170[a]
	82[b]	–	82[b]
Grass (UK)	267	–	267
Maize (USA)	216	100	116
Sugar cane (Hawaii)	890	284	606
Oil palm (Malaysia)	245	180	65

[a] Annual increment of closed crop canopy (photosynthetic).
[b] Mean annual yield (planting minus harvesting).
   (Adapted from ref. 35.)

TABLE 28
Energy Output of Cereal Straw and Timber Residues in UK

Crops	Dry weight (10^6 tonnes)	Energy (10^9 MJ)
Cereal straw: Currently burnt	3·5	64
Total	9·3	171
Timber: Harvested	1·5	27·5
Residue during harvesting	0·5	9·2
Residue from processing	0·3	5.5
Total residues	0·8	14·7
National energy consumption (1973)		9 260
Agricultural energy consumption (1973)		246[a]

[a] Used from growing crops (includes those for fertilisers, agricultural machinery, etc.).
(Source: ref. 35.)

used in some other ways in the UK and some other countries. For the rest of the world, in particular the tropical and sub-tropical regions, cellulose forms a very attractive path to energy solutions. In the developing world presently the use of fuelwood is the main source of energy. Uganda, Nigeria, Tanzania, Malawi and Nepal are a few nations that are heavily dependent on fuelwoods. Food crops are also abundant in these countries and considerable wastes accrue from these crops and timbers during planting and harvesting periods. Earl[36] constructed Table 29 which shows the variation of GNP with fuelwood consumptions for various countries. In Tables 30–32 Earl presents the world's renewable forest energy resources and the utilisation of the world's incremental forest energy resources as of 1970. Cooper,[35] Earl[36] and Leach[37] have published extensively on food production, crop growth and agriculture in general in relation to energy utilisations and outputs. In particular, Leach[37] gives the energy input–output ratios for many crops in various countries of the world.

From these tables one sees that the world as a whole can yet gain much from the use of solar energy via photosynthesis. The cultivation of 'energy crops and woods' could be made national programmes by many countries. Oil and gaseous products from such cellulosic materials treatments can help to maintain the present fossil fuel consumption rates. While technological progress can shape the trend of events, while the nuclear-oriented

TABLE 29
Per Capita GNP and Energy Consumption for some Selected Countries

Country	GNP per capita (US$)	Consumption per capita fuelwood (m³)	Energy consumption per capita (kilograms coal equivalent)		% of total energy supplied by fuelwood
			Forest only	Total	
Malawi	80	0·77	335	376	89·1
Nepal	80	0·57	248	259	95·8
Tanzania	100	2·30	999	1 042	96·0
India	110	0·19	83	274	30·3
Sri Lanka	110	0·31	135	291	46·4
Guinea	120	0·50	217	314	69·1
Nigeria	120	1·00	435	480	90·6
Malagasy	130	0·52	240	304	78·9
Uganda	130	1·07	478	531	90·2
Kenya	150	0·69	299	447	66·9
Rhodesia	280	0·63	274	838	32·7
Algeria	300	0·02	9	479	1·9
Ivory Coast	310	1·01	438	618	70·9
Zambia	400	0·90	391	900	43·4
Brazil	420	1·60	695	1 176	59·1
Cuba	530	0·20	87	1 140	7·6
Venezuela	980	0·63	274	2 427	11·3
Greece	1 090	0·25	109	1 259	8·7
Italy	1 760	0·14	61	2 492	2·4
Libya	1 770	0·20	87	569	5·3
USSR	1 790	0·36	157	4 356	3·6
UK	2 270	0·01	4	5 143	0·1
Belgium	2 720	0·20	9	5 438	0·2
West Germany	2 930	0·30	13	4 836	0·3
France	3 100	0·12	52	3 570	1·5
Canada	3 700	0·20	87	8 881	1·0
Sweden	4 040	0·41	178	5 946	3·0
USA	4 760	0·10	43	10 817	0·4

(Source: ref. 36. Based on data from UN (1973), IBRD (1972), FAO (1972).)

countries may want to investigate this field further, it is certain that a cheap domestic source of energy may presently be derived from waste. Developing countries must turn to cellulose as an important point to start while the developed countries must view cellulose as a source of continuously renewable energy.

TABLE 30
Fuelwood: Past, Present and Projected Consumption and Production (10^6 m^3)

Countries	1962	1975	1985
Developed countries	383	318	267
Developing countries	634	718	797
World total	1 017	1 036	1 064

(Source: ref. 36.)

TABLE 31
Fuelwood: Past, Present and Projected Consumption and Production (10^6 m^3)

Region	1960	1970	1975	1980	Change 1960–70 (1960 = 100)
Europe	91	62	74	58	68
USSR	98	87	80	82	89
North America	48	19	34	14	40
Latin America	188	223	220	244	119
Africa	208	255	246	312	123
Asia Pacific	397	475	545	626	120
World	1 030	1 120	1 199	1 335	109

(Source: ref. 36.)

TABLE 32
Fuelwood: Past, Present and Projected *Per Capita* Consumption (m^3)

Region	1960	1970	1975	1980	Change 1960–70 (1960 = 100)
Europe	0·20	0·13	0·12	0·10	65
USSR	0·45	0·36	0·31	0·29	80
North America	0·24	0·09	0·14	0·05	38
Latin America	0·88	0·80	0·70	0·72	91
Africa	0·75	0·73	0·63	0·66	97
Asia Pacific	0·24	0·23	0·25	0·27	96
World	0·34	0·31	0·30	0·30	91

(Source: ref. 36.)

4.3.1.2 *Just how attractive is cellulose as an energy carrier?*

Though this fundamental question seems to question the validity of the last section it is pertinent that it should be asked at this stage. The data given above are based on projections and literature works. As in most projects practical experience may differ from theoretical forecasts. As in the case of the Holden Town study most municipal wastes constitute 20–40% cellulosic materials, about 25% non-combustibles and the rest some other forms of combustibles. In less developed countries the proportion of cellulose in the wastes can go as high as 75%. In specially grown crops the percentage of water in the fresh crops can be as high as 75–80% in grass and about 10–50% in various grains.[37] This indicates quite low organic components in both the wastes and grown crops. But Cooper[35] shows that the growth rate of crops, aided by various forms of fertilisers, can help to overcome the water problem. Table 33 shows the growth rate of crops in

TABLE 33
Short-term Crop Growth Rates in Temperate and Tropical Environments

Environment and crop	Typical country	Crop growth rate	
		grams per metre²-day	kJ per metre²-day
Temperate			
Rye grass	UK	28	498
Red clover	New Zealand	23	409
Sugar beet	UK	31	552
Kale	UK	21	374
Wheat	Netherlands	18	320
Peas	Netherlands	20	356
Maize	UK	24	427
	New Zealand	29	516
Sub-tropical			
Lucerne	California (US)	23	409
Cotton	Georgia (US)	27	481
Rice	South Australia	23	409
Maize	California (US)	52	926
Tropical			
Cassava	Malaysia	18	320
Rice	Philippines	27	481
Napier grass	El Salvador	39	694
Sugar cane	Hawaii	37	659

(Adapted from ref. 35.)

TABLE 34
Annual Production of Crops in Temperate and Tropical Environments

Crop	Country	Yield crop components	Dry matter (tonnes per hectare)	Growing season (days)
Forage crops				
Perennial rye grass	UK	Forage	25	365
	Netherlands	Forage	22	365
Sorghum	California (US)	Forage	47	365
Napier grass	Puerto Rico	Forage	85	365
Plantation crops				
Sugar cane	Hawaii	Total	64	365
		Sugar	22	365
Oil palm	Malaysia	Total	40	365
		Fruit	11	365
		Oil	5	365
Cereals				
Wheat	Netherlands	Grain	6	153
Barley	UK	Grain	6	148
Rice	Japan	Grain	7	123
	Philippines	Grain	10	122
	Peru	Total	22	205
		Grain	12	205
Maize	UK	Grain	5	160
	Iowa (US)	Grain	9	141
	California (US)	Total	26	131
		Grain	13	131
Roots and tubers				
Sugar beet	UK	Total	23	217
		Roots	14	217
		Sugar	8	217
	Washington (US)	Total	32	230
		Roots	26	230
		Sugar	14	230

(Adapted from ref. 35.)

the temperate and tropical regions, while Table 34 shows the annual growth forecasts for 1975 of some crops.

Thus in quantifying the benefits of 'energy crops and wastes' one has to quantify the energy contributions, energy demands and hence costs estimates of processes dealing with cellulose (crops) conversion to oil and gases. And in quantifying the energy demands of such processes the parameters to look for are the energy to dry fresh crops or crop wastes, the energy inputs for the growth of these crops, the energy of the crops themselves, the energy for the collection of the crops and finally the energy

TABLE 35
Support Energy Inputs for Grass and Legume Production

Component input	Perennial ryegrass (MJ per hectare-year)	Lucerne (MJ per hectare-year)
Fertiliser N[a]	25 584	0
Fertiliser P[b]	2 263	1 364
Fertiliser K[c]	1 796	2 499
Herbicides	77	39
Fuel for field operations	5 941	5 941
Total support energy	35 661	9 843
Output at 10 tonnes dry material per hectare	177 820	177 820

[a] Nitrogen
[b] Phosphate
[c] Potash
(Source: ref. 35.)

TABLE 36
Support Energies for some Crops

Crop	Country	Type of labour input	Support energy (MJ per tonne-year)
Maize grain	UK 1973–4	Mechanised	5 253
Sugar beet	UK 1968–72	Mechanised	2 733
Grass (low efficiency)[a]	UK	Mechanised	2 427
Grass (high efficiency)[a]	UK	Mechanised	5 240
Corn	Guatemala	Axe and hoe	1 136
Corn	Guatemala	Mechanised	3 912
Corn	Mexico	Axe and hoe	505
Corn	Mexico	Mechanised	3 171
Corn	Nigeria	Axe and hoe	1 473
Corn	Philippines	Mechanised	3 051

[a] Low and high efficiency fertilisers used for growing crop.

input to the final processing of the organics to products. The 'support energy', required for crop growth will include energy for the synthesis of fertilisers, energy input to operations in the field such as tractors and other machinery, and energy input to herbicides and pesticides that may be used in the field. Quite often crop drying could be done, in the tropical and subtropical regions, by exposure of crops to sunlight. But in temperate regions some other forms of drying techniques have to be used. Energy used in these techniques has to be accounted for. Collection of dry materials can be done in the same way as in the case of municipal refuse—by trucks of 20–50 m³ capacity.

From Tables 35 and 36 it can be seen that grass, sugar cane and sugar beet, and palm oil have about the fastest growth rates. In the following sections grass is taken as the crop for case study though any other crop can be used.

4.3.2 Energy requirements of specially grown crops conversion to oil: 'From grass to grace?'

4.3.2.1 Support energy
This is the energy input required to make the crops grow at an economically acceptable rate, i.e. 'catalytic energy' to speed the growth rate of the crops. Appendix B3 deals with the evaluations of support energy which, for grass, is taken as 3850 MJ per tonne-year of raw crop.

TABLE 37
Fuel Energy Inputs for Four Vehicle Sizes, UK 1968

Conditions of use of vehicle	Minibus	7 tonnes truck	12 tonnes Leyland truck	20 tonnes 'Mandator' + trailer
Unladen weight (tonnes)	1·24	4·48	7·45	7·71
Half load weight (tonnes)	–	2·41	6·53	13·03
Full load weight (tonnes)	0·396	5·56	13·03	20·64
Fuel (MJ per vehicle kilometre)				
Rural roads: empty	4·33	8·31	8·66	12·1
half load	–	8·92	10·7	14·8
full load	4·76	9·83	11·9	15·3
City centre: empty	6·73	9·09	12·0	17·1
half load	–	12·3	16·9	23·7
full load	7·32	14·5	21·2	26·9

(Source: ref. 37.)

4.3.2.2 Gathering (collection) energy

Most food materials travel by road in the UK. Thus the most likely mode of collection of these specially grown crops, for conversion purposes, will be by road. Appendix B4 deals with the energy calculations of this operation; see, for example, Table 37. Energy in the form of labour is negligible in this analysis (by convention).

4.3.2.3 Total energy input for entire process

Basically the energy inputs will be divided into three subsections, viz. support energy input, collection energy input and processing energy input. The last corresponds to the energy input required for converting grown crops and crop wastes to oil.

In calculating the total energy requirements the assumptions are, first, that the conversion energy inputs will be patterned to those of the Holden refuse plant and, secondly, that drying is done in the fields by the sun. Improved technologies and sensitivity studies may improve upon these assumptions, but for now they are accepted. Based on a 1000 tonnes daily capacity of plant (Appendix B5) the energy inputs given in Table 38 are calculated.

TABLE 38

Inputs	Energy requirements (10^6 MJ per day)
Electricity for plant operations, e.g. grinding	1·420[a]
Plant and capital equipment	0·110[a]
Catalyst and chemicals	0·098[a]
Furnace and boiler feed gas	2·300[a]
Hydrogen production	0·410[a]
Crop material (heat value)	19·540[b]
Support energy	0·012[c]
Collection and transport	0·007[c]
Total	24·00

[a] Based on Holden refuse plant design.
[b] Average heating value of dry crops taken as $19·54 \times 10^3$ MJ per tonne.
[c] From Appendix B5.

4.3.2.4 *Process outputs and NER calculations*

From the stoichiometry of cellulose hydrogenation to oil and gases*

$$4\,C_6H_{10}O_5 + 12\,H_2 \longrightarrow C_6H_{14} + C_{12}H_{26} + 4\,CO + 2\,CO_2 + 12\,H_2O$$

it can be seen that theoretically about 648 tonnes of cellulose should react with 24 tonnes of hydrogen to give about 256 tonnes of hydrocarbon products. This is a theoretical yield of about 2·76 barrels of oil equivalent per tonne of cellulose treated. However, there are losses in the process and the light hydrocarbons and gases are used up in the process itself. Therefore another assumption to be made here is that products will be similar to those of refuse treatment. This makes it possible to apportion 83% of the GER of the process to oil production. On the basis of 2·15 barrels of oil per tonne of crop, a 1000 tonnes per day plant should yield 360 tonnes of oil. Thus

$$\text{GER for oil product} = \frac{1}{360} \times 24\!\cdot\!0 \times 10^6 \times 0\!\cdot\!83$$

$$= 55\,333\,\text{MJ(th) per tonne of oil}$$

The calorific value of the oil is 43 000 MJ(th) per tonne of oil. Therefore

$$\text{NER} = \text{GER} - \text{CV}$$
$$= 55\,333 - 43\,000$$
$$= 12\,333\,\text{MJ(th) per tonne of oil}$$

Hence

$$X_f \text{ value} = \frac{\text{NER}}{\text{CV}}$$
$$= \frac{12\,333}{43\,000}$$
$$= 0\!\cdot\!287$$

Analysis based on a 2000 tonnes per day plant gave an X_f value of 0·280. It should be noted here that the emphasis is on an oil product, since this section is devoted to production of synthetic oils from various other sources. Cost estimates of cellulose-based processes are dealt with later in Part II.

*A more representative equation than that in Section 4.1.

CHAPTER 5

Pyrolysis of Refuse to Synfuels

5.1 PREAMBLE

Although this is also a cellulose-based process the chemistry of it is different, pyrolysis being the chemical decomposition of matter without oxidation. In this case the matter is municipal wastes with a high percentage of cellulosic materials. The process involves heating the material at atmospheric pressure in the absence of air, and this basically differentiates it from direct incineration. Pyrolysis has the advantages of low pressure operations and no need for hydrogen and catalysts, and hence relatively cheaper operational costs compared to some other processes. However, it has the disadvantage of variability of output products, typically low Btu gas, usually not suitable as pipeline gas, char with much sulphur contamination and heavy tar-like oil. However, the products in pyrolysis processes can be predetermined and the pyrolytic reactor tailored to maximise the yield of product. Much work has been done on this aspect of conversion processes though no reported work on detailed economic evaluation has been published. Thus an attempt will be made here to estimate some costs. Many of the published data on costs of pyrolytic reactors are based on input refuse capacity, i.e. costs per tonne of refuse treated each day. This is reasonable since it has been noted earlier that the products are generally low in quality compared to those of other processes such as hydrogenative extraction of coal. However, as fossil fuels such as petroleum and gases become scarcer there may be need to further treat these 'quasi-useless' products to improve their qualities. Bearing this in mind, and noting that refuse and plants may play some role in energy economy in future, it will be worth while to quantify costs per unit tonne of outputs, e.g. £ per tonne of output heavy oil. In this section of Part I only the oil product will be considered. Chapter 11 in Part II will treat costs of gas produced. The char may be neglected as of now. Again in trying to cost these products the sociological factors such as taxation, etc., will be neglected.

5.2 THE SAN DIEGO GARRET PLANT

Many research groups are investigating this process of pyrolysis but currently only three variants of the process are fully developed, viz. the Garret process, the Monsanto process and the US Bureau of Mines pilot plant. The Monsanto process, developed by Monsanto Enviro-Chem Systems Inc., is currently at a commercial scale level, processing 1000 tonnes of refuse per day and producing gas and char. The BuMines pilot plant produces oil, char and gases. The Garret process also produces char, gas and oil but mainly with a view to maximising oil production. For detailed information on these reference is made to 'Energy Alternatives',[28] Flanagan[38] and Levy.[39]

The San Diego 200 tonnes per day process will be discussed here. Levy[39] gives a comprehensive description of the process. In summary, the process involves two stages of shredding, the primary stage reducing feed to a nominal size of 7·5 cm and the secondary stage reducing dried particles ex-primary shredder to 0·32 cm maximum size. After the non-combustible materials have been separated from the combustibles the latter are fed to the pyrolytic reactors where they come in contact with recycled char at 760 °C (1400 °F). The reaction is usually endothermic at this range of temperature and favours predominantly the production of gas. However, at the exit temperature of 510 °C (950 °F) the reaction is slightly exothermic and favours the production of oil. Thus to maximise oil production, conditions have to be adjusted. Figure 20 is a scheme of a pyrolysis process and Fig. 18 gives the material balance of a 200 tonnes typical San Diego plant. The refuse composition of San Diego is approximately 25% moisture content, 52% organics, 13% metal and 7% glass. The rest are miscellaneous solids. Thus for a throughput of 200 tonnes each day, 150 tonnes are combustibles, from which the products obtainable are 34 tonnes of liquid fuel, 24 tonnes of ferrous metals, and 10 tonnes of glass.

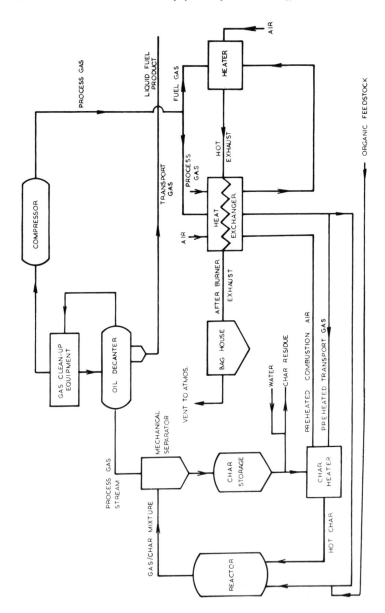

Fig. 20. Pyrolysis schematic. (Source: ref. 39.)

TABLE 39

Input operations	Energy requirements $(10^6 \ MJ(th) \ per \ day)$
Electricity for shredding, etc.	0·323
Plant and capital equipment	0·091
Refuse (heat value)	1·940
Oil for heating	0·016
Refuse collection[a]	—
Total	2·370

[a] Treated as labour cost element—Part II.

5.3 ENERGY REQUIREMENTS IN GARRET PROCESS

5.3.1 Gross energy requirements

In Appendix B6 the calculations of the energy requirements of the process are carried out. These are based on a 200 tonnes per day San Diego plant and the results are summarised in Table 39.

The gross energy requirement of the process is

$$2·37 \times 10^6 \ MJ \ per \ day$$

5.3.2 Process outputs and NER calculations

From Fig. 18 the feed of 150 tonnes of dry combustible refuse gives 33·8 tonnes of liquid fuel, 27·4 tonnes of gas and 11·1 tonnes of char. An assumption is made here in energy apportioning. Since the non-combustibles such as glass and metal will be taken off and sold for credits, and their prices are subject to sociological factors, they will be neglected. However, in the capital cost element, they will be considered and a portion of capital costs of plant attributed to them.

The products are

   (i) oil—33·8 tonnes per day (CV = 24 400 J per gram);
  (ii) char—11·1 tonnes per day (CV = 21 000 J per gram);
 (iii) gas—27·4 tonnes per day or 17×10^6 litres per day (CV = 20 000 J per litre).

% of GER attributable

$$\text{to oil product} = \frac{33\cdot8 \times 24\,400 \times 10^6}{(33\cdot8 \times 24\,400 \times 10^6) + (11\cdot1 \times 21\,000 \times 10^6) + (17 \times 20\,000 \times 10^6)}$$

$$= 59\%$$

and

$$\text{GER for product oil} = \frac{1}{33\cdot8} \times 2\cdot37 \times 10^6 \times 0\cdot59$$
$$= 41\,370 \text{ MJ per tonne}$$

The calorific value of oil is 24 400 MJ per tonne. Therefore

$$\text{NER of oil} = \text{GER} - \text{CV}$$
$$= 41\,370 - 24\,400$$
$$= 16\,970 \text{ MJ per tonne}$$

TABLE 40
Energy Requirements for Plants of Capacities Greater than 200 tonnes per day

Process inputs	Energy requirements (10^6 MJ per day)	
	Plant capacity of 1000 tonnes of refuse per day	Plant capacity of 2000 tonnes of refuse per day
Electricity	1·615	3·23
Plant equipment	0·238	0·361
Refuse input (heat value)	9·70	19·40
Heat for burners, etc.	0·08	0·16
Total plant GER	11·63	23·15
GER of product oil per tonne	0·040 6	0·040 6
Calorific value of oil per tonne	0·024 4	0·024 4
NER of product oil per tonne	0·016 2	0·016 0
X_f value of oil	0·664	0·656

and

$$X_f \text{ value of oil} = \frac{\text{NER}}{\text{CV}}$$
$$= \frac{16\,970}{24\,400}$$
$$= 0.70$$

As a matter of interest, calculations were carried out for capacities of plants greater than 200 tonnes per day, and in fact a 2000 tonnes capacity plant may be built at Bridgeport. The San Diego plant is aimed at proving conclusively that pyrolysis has both technical and economical viability.[38] It is also aimed at gaining operating experience before any scale-up plant is considered. The levels of product yield in a plant such as the San Diego plant are 1·2 barrels of heavy oil per tonne of refuse pyrolysed, 0·60 tonnes of char per tonne of refuse and 85 000 litres of gas (atmospheric pressure) per tonne of refuse.

To be of any commercial importance scale-up calculations are necessary. The 2000 tonnes per day plant at Bridgeport may just be the solution. In calculating costs for such a plant the following assumptions were made: the six-tenths rule was used for costs of equipment; electrical requirements for grinding, etc., were assumed to be at the same rate, i.e. 136 kWh per tonne of refuse processed; utilities costs were assumed to vary directly with plant capacity. Appendix B7 deals with sample calculations for a 2000 tonnes per day plant and results for energy requirements of 1000 tonnes per day and 2000 tonnes per day are given in Table 40.

CHAPTER 6

Conversion of Cellulose to
Gaseous Fuels
by Biological Digestion

6.1 PREAMBLE

Coal and cellulose form the major sources of synthetic gas production amongst the fuels considered in Chapters 1–5. Tar sands and oil shales give predominantly liquid oil products.

Solar synthetic natural gas (SNG) production by anaerobic digestion of specially grown plants is a rapidly growing field of research in the US and other countries concerned with solar energy collection via photosynthesis. Specially grown plants vary from common rye grass and napier grass to fuelwood. Reports have also been published on methane gas production from refuse dumps.[40] The rubbish is believed to be digested by micro-organisms present in the refuse giving off methane gas. Though this process of methane collection from refuse is quite feasible in terms of yields, the associated problems in operation and the exorbitant costs involved have made the process not too attractive. However, with specially grown crops digested in specially built reactors the story can be quite different. A number of investigators, viz. Graham et al.[41], Fraser[42] and Sitton and Gaddy,[29] have carried out extensive preliminary calculations on plants as energy carriers and have engaged in some economic evaluations of such processes. Sitton and Gaddy[29] reported processes based on crop wastes and fuel crops giving returns of about 19–27% annually. Fraser projected a process with a net potential of SNG product of 14×10^{12} scf annually. Average cost of producing SNG from such a process is reported as approximately $ 5·00 per 1000 scf (1974 price levels). Graham et al.[41] estimated that, at 1976 price levels, the costs of producing fuels from crop wastes and grown crops are of the following order: for crop wastes, pyrolytic reactors can yield fuels at a cost of $ 0·73 per 10^9 J of energy equivalent product; for specially grown crops the costs could be as high as $ 4·83 per 10^9 J; with the fermentation process (bioconversion) estimated costs of production are $ 2·62 per 10^9 J of methane yield for crop wastes and $ 6·72 for specially grown crops.

The attraction of such fermentation processes lies in the low investment profile and the high energy potential yield. A process typical of that proposed by Fraser[42] has a net yield of 14×10^{12} scf daily which, at a heating value of about 1·054 MJ per scf should provide the US with an inexhaustible energy source valued at $14·8 \times 10^{12}$ MJ. This is slightly above half of the total annual natural gas consumption in the US. But investment levels are at about $ 3·00 per daily scf of capacity. The Sitton and Gaddy proposed process has an estimated 50·0 million cubic feet of methane production daily. Total investments are about $ 45·7 million. This gives an investment level of about $ 1·0 per daily scf of capacity. However, the problem with most processes based on specially grown crops would be the question of land.

For a country like Britain land availability is bound to be a primary factor to consider. In the US the problem may be more relaxed and in some developing countries, such as Nigeria, this should not cause any concern. The energy approach to cost estimate will be used here to treat two case studies.

6.2 THE PRINCIPLE OF BIOCONVERSION PROCESSES

The process of obtaining methane from plant materials involves the grinding of the feedstock (mainly cellulose) and the biological digestion of the ground matter in reactors. Usually, after grinding, the biomass is mixed with water and made into slurries of about 10% solids concentration. The reactions involving the conversion of biomass to methane and related gases are usually slow and could be approximated by first order kinetics. Sitton and Gaddy[29] gave an example of reaction rate constant of 0·2 per day for a reaction involving leafy materials such as grass. The cellulose is first converted to organic acids and alcohols by micro-organisms; these intermediate products then break down further to yield methane, CO_2 and some unwanted gases. The CO_2 is scrubbed off the methane gas and the latter is passed on for use as either pipeline gas or petrochemical feedstock. In the example just cited about 50 million cubic feet per day of methane were

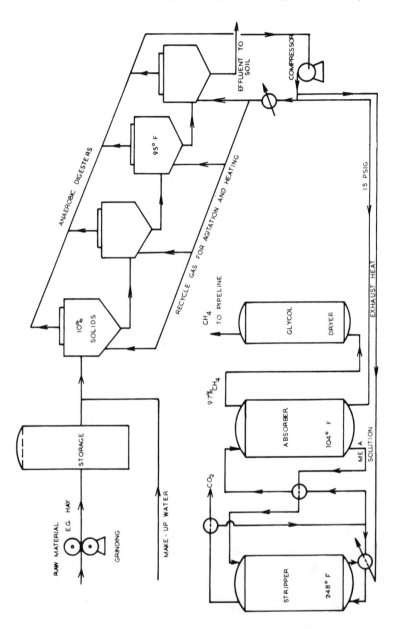

Fig. 21. Bioconversion process for producing methane. (Source: ref. 29.)

obtained by treating 4460 tonnes of biomass per day. Reactor volumes were as large as 5 million gallons capacity.

6.3 CASE 1: METHANE FROM LEAFY CROP MATERIALS BY BIOLOGICAL DIGESTION

Figure 21 is a schematic of a process presented by Sitton and Gaddy.[29] Raw materials (either crop wastes or specially grown crops) are field cut, baled into about 1 tonne bales and transported to a stockpile at the plant. These are ground by a hammer mill, passed through a shredder and made into 10% solid slurry. Reactors of about 5 million gallons capacity are used in the digestion of the biomass; the reactors are insulated steel tanks and are usually operated in series. Reaction conditions are about 35 °C (95 °F) and 15 psig pressure. The micro-organisms then anaerobically digest the crop matters in the slurry. The effluent gases are collected and passed through drying and cleansing stages. A proportion of the gases is recycled into the reactors for the purpose of agitation and heating; the effluent sludge (the undigested portion of the feed) is disposed of. Conversion efficiencies are about 94% on a thermal basis and the yield of methane is estimated at 1215·0 litres per kilogram of carbon destroyed. In this design study hay, cornstalks and oak leaves were used as raw materials. Typical carbon content of these plants is given as 35–40%, and with about 80% carbon destruction during digestion the estimated yield of methane in the studies was on average 343 litres per kilogram of dry crop matter or 12 320 scf of methane per tonne of dry crop matter treated.

6.4 CASE 2: METHANE FROM WOODY PLANT MATTER BY ANAEROBIC DIGESTION

Sitton and Gaddy[29] concentrated on leafy materials while Fraser[42] carried out a bioconversion process based on woody materials as feedstock. In his design green deciduous wood chips are first fed by means of a rotary valve into a double-revolving disk attrition mill pressurised to about 180 psia with steam from the steeping tank (Fig. 22). The wood chips are assumed to contain about 33% moisture on a total weight basis and grinding energy requirements are about 1·00 MJ(e) per kilogram of wood. After grinding, the wood particles are dropped into a steeping tank where these particles are made into slurries, which are then passed to the digesters. The processes

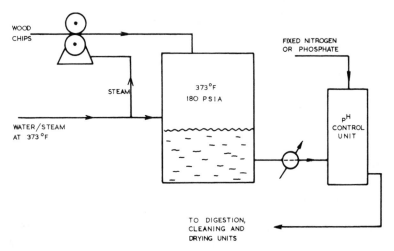

Fig. 22. Diagrammatic representation of methane production from woody plant materials.

TABLE 41
Chemical Composition of Average Hardwood

Wood components	'Average' bone-dry hardwood (weight %)	'Average' bone-dry softwood (weight %)
Hemicellulose, five-carbon polysaccharides	20	10
Hemicellulose, six-carbon polysaccharides	5	15
Cellulose	45	42
Lignin	23	30
Acetyl groups	6	2·6
Ash	1	0·4

(Source: ref. 42.)

of digesting, drying and cleaning should be quite similar to those of Sitton and Gaddy. Yield with wood chips is about 10 680 scf of methane per tonne of dry deciduous woody plant material. Fraser[42] gives the chemical composition of wood in Table 41 and the estimated yield of methane per pound of wood in Table 42.

The most important function of the steeping tank is to allow time, and to provide conditions, for promoting partial solubilisation of the ground

TABLE 42
Theoretical Yield of Methane from Anaerobically Digestible Materials

Digestible materials	scf per pound of bone-dry hardwood	scf per pound of bone-dry softwood
Methane from five-carbon polysaccharides	1·44	0·72
Methane from six-carbon polysaccharides	0·35	1·05
Methane from cellulose	3·16	2·95
Methane from acetyl groups	0·38	0·16

(Source: ref. 42.)

woody material. At a steamed temperature of 190 °C (373 °F) it is estimated that about 1% of the woody material will be solubilised in 5 min. When solubilisation is complete the slurry is cooled and the pH adjusted to about 7·5–8·0 by the addition of fixed nitrogen or phosphate, and the slurry is then pumped to the digesters. Solid contents are known to be about 10–12%.

In the digesters about 93% of the cellulose, all of the hemicelluloses and all of the acetyl groups are digested. Reaction conditions are about 60 °C (140 °F) and just above atmospheric pressure. The evolved gases, 60% methane and 40% carbon dioxide, are compressed to about 1000 psia and passed to drying and cleaning stages.

6.5 ENERGY REQUIREMENTS IN BIOCONVERSION PROCESSES

6.5.1 Case 1: Leafy matter
Calculations of this process are shown in Appendix B8 and the results, based on 4380 tonnes of crop matter per day, are summarised in Table 43.

TABLE 43

Input operations	Energy requirements (10^6 MJ per day)
Plant and capital equipment	0·54
Crop material (heat value)	67·30
Power for grinding, heat, etc.	3·92
Crop collection	0·032
Total	71·710

The total GER of the process is

$$71.71 \times 10^6 \text{ MJ per day}$$

6.5.1.1 Process outputs

No indication is made in the original design of the ratio of carbon dioxide to methane formed. But Hungate[43] indicates that the process of anaerobic digestion occurs at ordinary temperatures and pressures with a conversion efficiency as high as 94% on a thermal basis. Since CO_2 can be assumed to be non-contributory on a thermal basis the 94% can be assumed to be the yield of methane. Therefore

$$\text{GER of methane} = 71.70 \times 10^6 \times 0.94$$
$$= 67.4 \times 10^6 \text{ MJ per day}$$

The calorific value of methane is 1.054 MJ per scf of gas.[30] Thus

calorific value of 50×10^6 csf of methane $= 52.7 \times 10^6$ MJ per day

and therefore

$$\text{NER of methane} = \text{GER} - \text{CV}$$
$$= (67.4 - 52.7) \times 10^6$$
$$= 14.7 \times 10^6 \text{ MJ per day}$$

$$X_f \text{ value} = \frac{\text{NER}}{\text{CV}}$$
$$= \frac{14.7 \times 10^6}{52.7 \times 10^6}$$
$$= 0.28$$

TABLE 44

Input operations	Energy requirements (10^6 MJ per day)
Plant and capital equipment	2.94
Crop matter (heat value)	148.00
Power for grinding, etc.	16.00
Water/steam for solubilisation	43.70
pH control chemicals	6.02
Crop collection	0.15
Support energy for plant growth	0.114
Total	217.0

6.5.2 Case 2: Woody matter

Calculations on this are shown in Appendix B9. The summary is given in Table 44, based on 9430 tonnes of wood per day.

The GER of total plant operations is

$$217 \times 10^6 \text{ MJ per day}$$

6.5.2.1 Process outputs

Fraser[42] estimates an output ratio of methane to carbon dioxide of 60 : 40. Therefore the fraction of the plant's GER attributable to methane production is 60%. The methane yield is 92·4 million scf per day. Hence

$$\text{GER of methane product} = 217 \times 10^6 \times 0·6$$
$$= 130·12 \times 10^6 \text{ MJ per day}$$

The calorific value of methane is 1·054 MJ per scf.[30] Therefore

$$\text{heat value of product} = 1·054 \times 92·4 \times 10^6$$
$$= 97·4 \times 10^6 \text{ MJ per day}$$

Thus

$$\text{NER of methane} = (130·12 - 97·4) \times 10^6$$
$$= 32·72 \times 10^6 \text{ MJ per day}$$

and

$$X_\text{f} \text{ value of process} = \frac{\text{NER (methane)}}{\text{CV (methane)}}$$
$$= \frac{32·72 \times 10^6}{97·4 \times 10^6}$$
$$= 0·336$$

CHAPTER 7

Production of Synthetic Gases from Coal

7.1 PREAMBLE

Coal is a very important future source of pipeline gas for domestic consumption, and of the many processes which can convert coal to synthetic gases the most important ones are:

(i) pyrolysis
(ii) gasification, and
(iii) the advanced combined cycle process being investigated by the Coal Research Centre at Cheltenham.

The pyrolysis process followed by the low pressure Kellogg version of char gasification has already been mentioned in Chapter 1. However, the energy requirements of the process, with respect to gaseous products, will be investigated further. The net energy requirements should be similar to those of the liquid fuel product. The direct gasification of coal and the combined cycle processes will not be treated here, as Massey[44] and Merrick[45] have projected technical details and cost estimates, respectively, on these processes. Also a volume of CEP contains reports on coal processing to both gaseous and liquid fuels.[46] A typical cost value quoted for the combined cycle process at the NCB is about £ (1978) 2·0 per 1000 scf of gas for coal price of about £ (1978) 22·60 per tonne. At the same coal price levels the Lurgi plant produces SNG at a cost of about £ (1978) 2·20 per 1000 scf of methane.

7.2 COED PROCESS (PYROLYSIS OF COAL PLUS CHAR GASIFICATION)

From Section 1.4.3 and Appendix A3 Shearer's design on coal pyrolysis followed by resulting char gasification[17] gave the following products: 27 275 barrels of syncrude with calorific value of 6120 MJ per barrel, 7·08 million cubic metres of pipeline gas with heat value of 34·31 MJ per cubic metre and 1900 barrels of light hydrocarbons with heat value of 4120 MJ per barrel. Other byproducts, e.g. phenol and sulphur, were also obtainable. These were based on a daily coal feed of 28 455 tonnes (heat value 26 040 MJ per tonne) and a total plant gross energy requirement of $7·605 \times 10^8$ MJ per standard day.

7.2.1 Apportioning of GER amongst products

As in the case of synthetic crude a fraction of the total GER can be apportioned to the pipeline gas product. This is given by the relation

% of total plant GER attributable to gas production

$$= \frac{7·08 \times 10^6 \times 34·31}{(7·08 \times 10^6 \times 34·31) + (27\ 275 \times 6120) + (1900 \times 4120) + (941 \times 400) + (36 \times 50\ 850)}$$

$$= 57·85\%$$

7.2.2 Calculation of NER and X_f value of product gas

The GER of product gas therefore is given as

$$7·605 \times 10^8 \times 0·5785 = 4·40 \times 10^8 \text{ MJ per daily production}$$

The heat value of daily gaseous product is given as

$$7·08 \times 10^6 \times 34·31 = 2·43 \times 10^8 \text{ MJ per day}$$

Therefore
NER of daily gas product

$$= \text{GER} - \text{CV}$$
$$= (4·40 - 2·43) \times 10^8$$
$$= 1·97 \times 10^8 \text{ MJ per day}$$

and

$$X_f \text{ value of product gas} = \frac{\text{NER}}{\text{CV}}$$
$$= \frac{1·9708 \times 10^8}{2·43 \times 10^8}$$
$$= 0·81 \text{ (as in the case of oil)}$$

Though no calculations are carried out on direct gasification of coal some indication will be given on the net recoverable fraction of such a process $(1 - X_f)$. Penner and Icerman[47] have detailed the Hygas–Electrothermal process of gasifying coal directly. In this process 16 265 tonnes per standard day are fed into the process. Also about 318 000 kW of electricity and steam are consumed by the process. Although no indication as to coal–steam ratio is given, a 1 : 2 ratio is most probable. Other energy inputs may be neglected in this evaluation since coal, steam and electricity form the major cost elements of the process. Outputs of the process are 250×10^6 scf of gas per standard day. Also about 3765 tonnes per standard day of char are obtained. Other byproducts include 900 tonnes per day of residue, 188 tonnes per day of CO_2 and 10 tonnes per day of hydrogen sulphide.

With the figures above, the total energy requirements of the process can be estimated as 590×10^6 MJ per standard day, and the proportion of total energy attributable to SNG production following the same pattern of distribution according to product heat values will be given as

$$\frac{250 \times 10^6 \times 0.975}{\text{(total product heat value)}}$$

where 0·975 MJ is the heat value of product gases per scf, heat value of char product is approximately 33 000 MJ per tonne, heat value of $CO_2 = 0$ and heat value of $H_2S = 19$ MJ per tonne. Therefore

$$\% \text{ of GER attributed to gases production} = \frac{2.438 \times 10^8}{3.66 \times 10^8}$$
$$= 66.64\%$$

i.e.

$$\text{GER of gases production} = 590 \times 10^6 \times 0.6664$$
$$= 3.932 \times 10^8 \text{ MJ per daily production}$$

Heat of daily gaseous products

$$= 250 \times 10^6 \times 0.975$$
$$= 2.438 \times 10^8 \text{ MJ}$$

i.e.

$$\text{NER of gaseous products} = \text{GER} - \text{CV}$$
$$= 1.494 \times 10^8 \text{ MJ per day}$$

and

$$X_f \text{ value of products} = \frac{1 \cdot 494 \times 10^8}{2 \cdot 438 \times 10^8}$$
$$= 0 \cdot 61$$

It can thus be seen that, from these rough calculations, the recoverable fraction of gas from coal gasified in terms of energy value $(1 - X_f)$ is in the same order as that of the pyrolysis – gasification process. An inferred conclusion here will be that any variances in costs of production of pipeline gases from either the pyrolysis process, the gasification process or the combined process will depend mainly on the capital investments of relevant process plants and labour employed in the process operations.

CHAPTER 8

Gaseous Fuels from Pyrolysis of Wastes

8.1 PREAMBLE

As noted in Section 5.2 the pyrolysis of refuse, and cellulosic materials in general, gives rise to gaseous products as well as liquid fuels and char. It was also noted that the char is heavily impregnated with sulphur and the gaseous products are of low energy value. However, the costs of producing the gaseous products and the energy requirements will be discussed, since for a completely integrated process it may be necessary in the future to treat these gases further and raise their heat values.

With the 200 tonnes per day San Diego plant it was noted that products obtainable are

  (i) oil—33·8 tonnes per day (CV = 24 400 J per gram);
  (ii) char—11·1 tonnes per day (CV = 21 000 J per gram);
  (iii) gas—17 million litres per day (CV = 20 000 J per litre).

From these the NER and X_f value of the gaseous products can be calculated.

8.2 NET ENERGY REQUIREMENTS AND X_f VALUE OF GASEOUS PRODUCTS

From the calorific values of products shown in Section 8.1, the percentage of total plant gross energy demand attributable to gas production can be calculated as

$$\frac{17 \times 20\ 000 \times 10^6}{(17 \times 20\ 000 \times 10^6) + (11 \cdot 1 \times 21\ 000 \times 10^6)} = 24 \cdot 32\%$$
$$+ (33 \cdot 8 \times 24\ 400 \times 10^6)$$

The gross energy requirement of the plant was also calculated as $2 \cdot 37 \times 10^6$ MJ per day. Therefore

$$\text{GER of gases} = \frac{2 \cdot 37 \times 10^6 \times 0 \cdot 2432}{17 \times 10^6}$$

$$= 0 \cdot 0339 \text{ MJ per litre of gas}$$

The heat value of gaseous products $= 0 \cdot 02$ MJ per litre. Therefore

$$\text{NER of gaseous products} = \text{GER} - \text{heat value}$$
$$= 0 \cdot 0139 \text{ per litre}$$

and hence

$$X_f \text{ value} = \frac{0 \cdot 0139}{0 \cdot 02}$$
$$= 0 \cdot 70 \text{ (as before)}$$

8.3 COMPARATIVE DISCUSSIONS OF NER AND X_f VALUES OF VARIOUS CONVERSION PROCESSES

Why engage in calculations of the net energy requirements (NER) and the net fractional recovery of converted fossil fuel sources $(1 - X_f)$ of the various processes? What significant part has energy demand to play in obtaining oils and gases from solid fuels? These questions refer us back to the introductory chapters. It has been shown that the costs of a product are bound to be dependent on the energy demands of the process producing it. It has also been shown that for a manufacturing process using 'priceless' feedstock to break even the following relation must hold:

$$P_f = \frac{\sum\limits_{j \neq f} P_j}{(1 - X_f)} \tag{3}$$

where P_f is the price of fuel used in the process and P_j are costs or prices of other inputs to the process which are not attributable to fuel inputs. $(1 - X_f)$ is the net energy yield of the process. To make more sense to the reader, let us view $(1 - X_f)$ as this: suppose 1 tonne of synthetic oil is to be made from cellulose by the pyrolysis method and 0·70 tonne equivalent of this product oil is used effectively to get the tonne of oil out. Then the net gain of the oil process is $(1 - 0·70)$ or 0·30 tonne of oil equivalent. The significance of this is that, before a conversion process is carried out, there is a need to quantify the gain of the process with respect to energy. This is necessary because of the impending depletion of present energy source reserves and the need to obtain alternatives. Just as inflation in monetary terms is becoming a matter of concern so also must energy scarcity and 'inflation' be considered. Processes are constrained by energy demands and it is positive thinking that processes with non-significant energy gains should be properly investigated before being commissioned. The denominator of the relation given above may be misleading. It does not imply that processes with X_f values greater than unity must have an infinite price of fuel to break even. It only implies that prices may soar very high. It is a practical experience that the Sasol process yields products that are just beginning to be competitive in price with products from other processes. In Part II the effect of X_f value on costs of products will be discussed.

TABLE 45

Processes considered	X_f values
Coal processes	
H-coal process to oil	0·602–0·627
Pyrolysis process to oil and char	0·466
Pyrolysis process to oil, gas and char	0·812
Hydrogenative extraction of coal	0·67–0·69
Supercritical extraction of coal to oil and char	0·32
Fischer–Tropsch gasification–synthesis to motor spirit	1·274
Fischer–Tropsch gasification–synthesis to methanol	0·792
Oil shale processes	
Recovery of oil from shale rocks	0·23 (average)
Oil sands processes	
Recovery of oil from tar sands	0·31–0·58
Cellulose processes	
Municipal refuse to oil	0·41
Grown plants to oil	0·29
Pyrolysis of refuse to oil and gases	0·70
Biological conversion of plants to methane gas	0·28–0·34

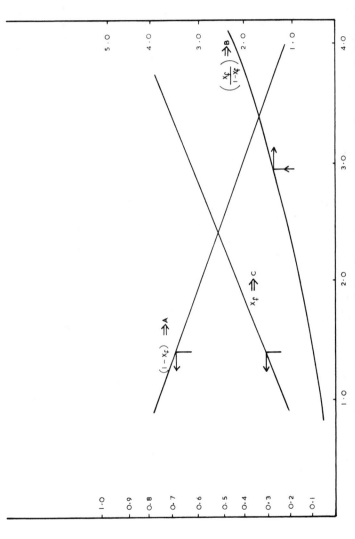

Fig. 23. Net energy requirements of conversion processes. Left ordinate: net energy recovery of processes $(1 - X_f)$; right ordinate: rates of energy demand-recovery of processes; abscissa: NER values of processes (10^4 MJ per tonne of product).

TABLE 46
Basic Parameters of some Conversion Processes

Conversion processes considered	NER of process (MJ per tonne of product)	Calorific value of product (MJ per tonne)	X_f value (fractional energy demand of process)	$1 - X_f$ value (fractional energy recovery of process)
Direct hydrogenation of coal to oil, H-coal process (Case 1)	18 912	42 700	0·443	NA[a]
H-coal process (Case 2)	25 690	42 700	0·602	NA
Pyrolysis of coal to oil and char mainly	19 930	42 800	0·466	NA
Pyrolysis of coal to oil and gasification of resulting char to gases	34 758	42 800	0·812	NA
Hydrogenative extraction of oil from coal (liquefaction)	31 844	46 830	0·680	NA
Supercritical extraction of coal	11 770	36 730	0·320	NA
Fischer – Tropsch's gasification – synthesis of motor spirit from coal	59 890	47 000	1·274	NA
Fischer – Tropsch's gasification – synthesis of methanol from coal	18 300	23 100	0·792	NA
Extraction of oil from tar sands	16 450	42 400	0·440	0·560
Extraction of oil from oil shale rocks	9 760	42 700	0·230	0·770
Liquefaction of municipal refuse to oil	17 700	43 000	0·410	0·590
Liquefaction of specially grown plants to oil	12 333	43 000	0·290	0·710
Biological digestion of crop matter to pipeline gas (methane)	16 200 (equivalent)	52 700 (equivalent)	0·310	0·690
Pyrolysis of refuse to oil and gases	17 280	24 840	0·700	0·300

[a] Not applicable. Subsystems considered in coal processes.

However, looking at all the conversion processes already discussed coal processes seem to have the highest values of X_f. As a recap X_f values are presented in Table 45. These figures definitely indicate that coal processes are mainly more energy demanding. Apart from the supercritical extraction process most other coal processes have X_f values greater than 0·40. For other fossil fuel processes X_f values are less than 0·40 except for the pyrolysis process and, perhaps, municipal refuse to oil.

From empirical calculations, and comparison of costs evaluated to present price levels (Part II), the following assertions will be made. For processes with X_f values greater than 0·60 cost estimates will be made with a linear three-factor model equation (see Part II), i.e. the cost of producing synthetic fuels from solid fossil fuels will be given by eqn. 1. For processes with X_f values less than or equal to 0·40 a first approximation can be made, namely $C = P_f$ and eqn. 11 of Part II will be used to determine costs of products ex the process. Between 0·40 and 0·60 the processes will be termed energetically 'non-definite', i.e. neither energy intensive nor non-intensive. Like any transition state, the costs can be evaluated within an error percentage limit of 20% with any of the above equations. This aspect is dealt with more fully in Part II. However, in summary, the molecular complexity of coal, the separation stages involved in coal processes, the thermal and pressure conditions required for coal processes to occur, and the use of catalysts make coal processes most energy demanding. Also, unlike the cellulose processes, electricity consumption, oxygen utility and hydrogen production contribute to the high energy demands of coal processes. Figure 23 shows the relationship between the net energy requirements of the various processes, their net energy recovery values $(1 - X_f)$, and $X_f/(1 - X_f)$, which are ratios of the net energy demands of the processes to their respective energy recoveries, and Table 46 tabulates these. The point of interception is of high significance and is discussed in Part II.

PART II

Energy Economics of
Synthetic Fuels Production—
Costs Estimates of Conversion Processes
and General Topics

CHAPTER 9

Estimated Costs of Producing
Synthetic Oil from Coal

9.1 PREAMBLE

The costs of producing a unit of output in any process can be expressed in terms of payments for fuels, labour, materials, transport, equipment, etc. These can then be expressed in terms of energy, labour and capital costs. The above statements presuppose that cost factors such as land, taxations, etc., which are highly dependent on sociological patterns, are not taken into account. In Fig. 1 the classification and factorisation of these costs elements by Chapman[3] are shown. The primary elements of cost considered are fuel energy inputs, labour inputs and capital investments. The last two represent the non-fuel factor of production that can be represented by personal income payments.

The cost of manufacturing products in industries can thus be presented as

$$C = \bar{X}_f \bar{P}_f + \sum_{j \neq f} P_j \tag{2a}$$

where $\bar{X}_f$ = total quantity of fuel input to process
    = gross energy requirements of product with respect to its calorific value—fractional for energetically viable processes;

  $\bar{P}_f$ = price of fuel used in process (as defined in Part I);

  $\bar{P}_j$ = other costs incurred in the process that are not due to energy inputs.

This approach to costing is to illustrate how the price of energy in any production process can relate to the total costs of any industrial product. In particular as fossil energy sources become depleted the need to evaluate the costs of alternative sources becomes more and more relevant to fixed prices of market commodities.

In the case of cellulose conversion processes the non-fuel costs are attributed to labour and equipment and plants capital costs. Since many sociologically oriented cost factors have been neglected the costs calculated here should be treated as 'guide-costs'. Even then the variances that might occur between budgets based on these costs and actual operational costs may turn out to be quite insignificant. This is due to the fact that the essential elements of costing have been taken as bases for these calculations. For a break-even situation to occur in such a manufacturing process, applying the energy analysis system, the cost of producing a product should be equal to the price of the fuel used to produce the product. In practical terms this implies that the process just pays off the non-fuel cost elements and the entire energy inputs to the process costed at the on-site value of the product. In the case of cellulose conversion to oil, for example, the total energy inputs should be costed at the on-site value of the product synthetic crude, i.e. $C = P_f$. With this statement eqn. (2a) becomes

$$C = X_f C + \sum_{j \neq f} P_j$$

or

$$P_f = X_f P_f + \sum_{j \neq f} P_j$$

where

$$P_f = \frac{\sum_{j \neq f} P_j}{(1 - X_f)} \tag{3}$$

i.e.

$$\text{price of product fuel} = \frac{\text{costs of non-fuel inputs}}{\text{net energy yield of process}}$$

The net energy yield $(1 - X_f)$ is similar to process efficiency. Suppose all energy input to the process is recovered at 100% efficiency; then $(1 - X_f) \longrightarrow 1$, and the price of product oil is then equal to costs of non-fuel inputs. However, most processes have energy efficiencies less than unity and hence the denominator $(1 - X_f)$. The difference between $(1 - X_f)$ and actual process efficiency is that $(1 - X_f)$ is in essence a recoverable fraction of the calorific value of the product while efficiency is usually defined as output/input ratio.

Using cellulose conversion processes as an example, again the cost of producing synthetic crude can be expressed as

$$C = X_f P_f + X_1 P_1 + X_c P_c \tag{1}$$

where $X_f =$ quantity of fuel energy input expressed as tonne energy equivalent of product syncrude;

$P_f =$ price of product syncrude;

$X_c =$ quantity of funds input (£ per tonne of syncrude product);

$P_c =$ price of funds, usually expressed as annual capital charges rate;

$P_1 =$ price of labour (£ per man-hour worked);

$X_1 =$ quantity of labour used in process (man-hours worked per tonne of product).

It should be noted that costing is on a 'per tonne of synthetic crude product' basis. For a process in which more than 85% of total fuel input is derived locally a first approximation could be made that the fuel input can be costed at the on-site value of product, i.e. $C = P_f$, in which case

$$C = X_f C + X_1 P_1 + X_c P_c$$

and

$$C = \frac{X_1 P_1 + X_c P_c}{1 - X_f} \tag{11}$$

It should be noted that if the labour costs to bring coal to the plant are rightly ascertained, $X_1 P_1$ then becomes the ascertained value, while the energy input value of coal can be costed at on-site value. But the usual practice with producers dealing with coal is to quote an all-embracing price of coal. This tends to invalidate the approximation made above. This aspect of the costing is dealt with later.

The denominator in eqn. (11) can be misleading since some processes have X_f values greater than unity and hence will tend to have infinite costs of production. In practical terms this is impossible. Processes having X_f values greater than unity essentially have high costs of production but are finite. In an attempt to eliminate a possible confusion based on the factor $(1 - X_f)$ plots of the variations of $(1 - X_f)$ and ratios of X_f to $(1 - X_f)$ with the NER of the various conversion processes are done here to demonstrate the optimum values of X_f for which eqn. (11) may be valid. Since in most conversion processes the calorific values of similar final products do not vary much the NER values are used for this analysis. As examples of the preceding statement, coal to oil processes give products of calorific values of about 42 700 MJ per tonne; plant matter to oil processes give products of calorific value of 43 000 MJ per tonne; plant matter to pipeline gases by biological digestion give products of calorific values of 52 700 MJ per tonne; coal to motor spirit (gasoline) gives products of calorific values of 46 830 MJ per tonne; while pyrolysis of wastes give low-Btu products with average heat values of 24 800 MJ per tonne. However, the NER of the various processes varies depending on the energy intensiveness of the process (see Table 43). The $X_f/(1 - X_f)$ factor determines the ratio of energy demands of the processes to their energy recoverables. At the interception point of plots A and B (see Fig. 23) $X_f/(1 - X_f)$ is equal to 1·50 and this gives a value of $X_f = 0·6$. Also the interception point of plots A and C gives $X_f = 0·50$. Essentially eqn. (11) gives cost values which are reliable with a high degree of accuracy for X_f values less than or equal to 0·40. For X_f values greater than 0·60 costs values obtained with eqn. (11) are highly susceptible to errors. The processes in the range $0·40 < X_f < 0·6$ may be termed 'non-energy definite', i.e. neither energy intensive nor non-intensive. In these cases costs can be calculated with either eqn. (1) or eqn. (11) with cost errors between ± 20% of actual values. The significance of these plots rests on the point that other fossil fuel sources may be developed

TABLE 47
(*f*) Values of Various Coal Processes

Process	Calorific values of coal used (MJ per tonne)	Main products output, calorific value (MJ per tonne)	(f) Value
H-coal (Case 1)	25 530	42 700	1·68
H-coal (Case 2)	18 100	42 700	2·36
COED (Case 1)	26 040	42 800	1·64
COED (Case 2)	31 400	43 000	1·37
CSF (Case 1)	27 170	46 830	1·72
CSF (Case 2)	27 360	46 830	1·71
Fischer–Tropsch (motor spirit)	20 600	47 000	2·28
Fischer–Tropsch (methanol)	20 600	23 100	1·12
NCB, hydrogenative extraction	32 950	43 210	1·31
NCB, supercritical extraction	34 130	36 726	1·07

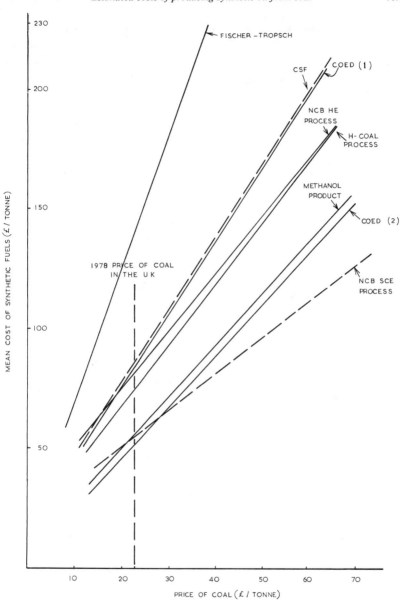

Fig. 24. *Cost of liquid synthetic fuel as a function of coal price and conversion process.*

in future with improving technology. What is being said is that the energy demand – recovery ratios should be estimated for all conversion processes. Hence, above X_f values of 0·6 the costs of production should be evaluated by eqn. (1). In this case the energy factor has to be redefined: $X_f P_f$ is redefined as $X_f P_f'(f)$ where X_f remains the same, P_f' becomes price of fuel feedstock, e.g. price of coal used (£ per tonne coal), and (f) is the ratio of total energy used, in coal equivalent, to the produced product (tonne equivalent coal per tonne oil). It should be noted that the figure of $X_f = 0·4$ given above, as optimum value for which eqn. (11) is highly applicable, is empirical and must be subjected to practical experiences and analyses. Suffice it here to say that a process using much energy and hence giving a demand – recovery ratio greater than unity is not attractive.

9.2 UNIT ENERGY COST ESTIMATE OF COAL PROCESS

Most coal processes have X_f values between 0·466 and 0·812, and for these eqn. (1) is more appropriate. The energy cost element is redefined as $X_f P_f'(f)$ as above. In Appendix C1 the values of (f) for all the coal processes considered are calculated, and these are presented in Table 47 and Fig. 24. The price of coal, P_f', varies between £ 12·83 per tonne (5 pence per therm) and £ 64·20 per tonne (25 pence per therm). The X_f values have been calculated already in Part I. The contributions of the energy unit costs of coal processes are shown in Table 48.

9.3 UNIT CAPITAL COSTS OF COAL PROCESSES PRODUCING OIL AND GASES

Appendix C2 deals with the evaluation of the capital costs per tonne of product of the coal conversion processes. The quantity $X_c P_c$ is calculated together as P_c represents the annual capital charge rate (in these processes taken as 15%) and X_c is the capital investment made for the processes. In some cases X_c represents initial investment while in other cases it represents total investment. In order to apportion these investments to the various products a factor D_c is introduced. This represents the fraction of the total products considered in terms of energy value. As an example, if a process produces oil (25 tonnes per day) of heat value 30 224 MJ per tonne and gas (17 million litres) of heat value 21 000 MJ per million litres, then the value D_c of oil product is given by the equation

TABLE 48

Estimated Cost of Producing Synthetic Oils from Coal (Three-factor Approach)

Process	Products capital costs $X_c P_c{}^a$ (£ per tonne)	Products labour costs $X_l P_l{}^b$ (£ per tonne)	Products energy factor $(X'_f)^c$	Ratio of CV of product oil to coal used $(f)^d$	Coal prices^e (£ per tonne)	Costs of producing syncrude (£ per tonne)
H-coal	12·00	0·39	1·443	1·68	12·83	47·30
(Case 1)	12·00	0·39	1·443	1·68	25·67	82·60
	12·00	0·39	1·443	1·68	38·50	117·20
	12·00	0·39	1·443	1·68	51·33	152·10
	12·00	0·39	1·443	1·68	64·20	187·00
H-coal	13·00	0·41	1·602	2·36	12·83	61·90
(Case 2)	13·00	0·41	1·602	2·36	25·67	110·40
	13·00	0·41	1·602	2·36	38·50	159·00
	13·00	0·41	1·602	2·36	51·33	207·40
	13·00	0·41	1·602	2·36	64·20	256·00
COED	16·80	0·60	1·812	1·64	12·83	55·60
(Case 1)	16·80	0·60	1·812	1·64	25·67	93·80
	16·80	0·60	1·812	1·64	38·50	132·00
	16·80	0·60	1·812	1·64	51·33	170·30
	16·80	0·60	1·812	1·64	64·20	208·60
COED	3·46	1·23	1·466	1·37	12·83	30·50
(Case 2)	3·46	1·23	1·466	1·37	25·67	56·20
	3·46	1·23	1·466	1·37	38·50	81·00
	3·46	1·23	1·466	1·37	51·33	108·00
	3·46	1·23	1·466	1·37	64·20	133·60
CSF	16·74	0·74	1·678	1·72	12·83	54·60
(Case 1)	16·74	0·74	1·678	1·72	25·67	91·72
	16·74	0·74	1·678	1·72	38·50	128·80
	16·74	0·74	1·678	1·72	51·33	166·00
	16·74	0·74	1·678	1·72	64·20	203·20
CSF	18·25	0·75	1·694	1·71	12·83	56·00
(Case 2)	18·25	0·75	1·694	1·71	25·67	93·40
	18·25	0·75	1·694	1·71	38·50	130·60
	18·25	0·75	1·694	1·71	51·33	168·00
	18·25	0·75	1·694	1·71	64·20	205·10

(Contd.)

TABLE 48 *(Contd.)*

Process	Products capital costs $X_c P_c{}^a$ (£ per tonne)	Products labour costs $X_l P_l{}^b$ (£ per tonne)	Products energy factor $(X'_f)^c$	Ratio of CV of product oil to coal used $(f)^d$	Coal prices^e (£ per tonne)	Costs of producing syncrude (£ per tonne)
Fischer–	26·00	0·96	2·274	2·28	12·83	93·50
Tropsch	26·00	0·96	2·274	2·28	25·67	160·00
(motor spirit)	26·00	0·96	2·274	2·28	38·50	226·60
	26·00	0·96	2·274	2·28	51·33	293·10
	26·00	0·96	2·274	2·28	64·20	360·00
Fischer–	10·00	0·41	1·792	1·12	12·83	36·20
Tropsch	10·00	0·41	1·792	1·12	25·67	61·80
(methanol)	10·00	0·41	1·792	1·12	38·50	87·80
	10·00	0·41	1·792	1·12	51·33	113·60
	10·00	0·41	1·792	1·12	64·20	139·40
NCB, hydro-	23·10	3·70	1·670	1·31	12·83	55·00
genative	23·10	3·70	1·670	1·31	25·67	83·00
extraction	23·10	3·70	1·670	1·31	38·50	110·00
	23·10	3·70	1·670	1·31	51·33	139·00
	23·10	3·70	1·670	1·31	64·20	167·20
NCB, super-	16·00	4·50	1·320	1·07	12·83	36·60
critical	16·00	4·50	1·320	1·07	25·67	56·80
extraction	16·00	4·50	1·320	1·07	38·50	75·00
	16·00	4·50	1·320	1·07	51·33	93·00
	16·00	4·50	1·320	1·07	64·20	111·10

^{a} From Table 49.
^{b} From Table 51.
^{c} $X'_f = (1 + X_f)$; X_f values are in Table 45.
^{d} From Table 47.
^{e} According to *Energy Trends*[53] the price of coal for the first quarter of 1978 in the UK was £22·6 per tonne.

$$D_c = \frac{25 \times 30\ 224}{(25 \times 30\ 224) + (17 \times 2100)}$$

$$= 0·68$$

Where there is only one product, $D_c = 1$. Table 49 gives the various values of $X_c P_c$ of the processes and their contributions to total costs of production are given in Table 48.

TABLE 49

$X_c P_c$ **Values of Various Coal Processes**

Process	Capital investments as given by ref. 11 ($ million)	Fraction of main product of total output (D_c value)	Products capital costs $X_c P_c$ values (£ (1978) per tonne)
H-coal (Case 1)	(1974)533	0·950	12·00
H-coal (Case 2)	(1974)533	0·960	13·00
COED (Case 1)	(1974)482·9	0·398	16·80
COED (Case 2)	(1970) 32·7	0·427	3·46
CSF (Case 1)	(1969)239·6	0·970	16·74
CSF (Case 2)	(1969)258·2	0·974	18·25
Fischer–Tropsch (motor spirit)	(1974)484·5	0·469	26·00
Fisher–Tropsch (methanol)	(1974)429·1	0·694	10·00
NCB, hydrogenative extraction	(1974) 75·6[a]	1·000	23·10[c]
NCB, supercritical extraction	(1974)140·1[b]	1·000	16·00[c]

[a] Presented by Davies *et al.*[20] as £ (1974)42 million.
[b] Presented by Maddocks and Gibson.[21]
[c] D_c value taken as unity in these processes.

9.4 UNIT LABOUR COSTS OF COAL PROCESSES PRODUCING OIL AND GASES

The price of labour, P_l, in these processes will be assumed to be similar to those presented by the *Digest of Energy Statistics*[48] for most industries in the UK.

For full-time employees the number of hours worked and average hourly earnings in April 1977 are as shown in Table 50. Thus the average weighted price of labour can be estimated, with a 10% rise to make for 1978, and a men–women ratio of 2 : 1 in employment, as

£ 1·85 per man-hour worked

X_l, the quantity of labour put into the process, will be assumed to be that for the conversion stage only. The price of coal employed as feedstock takes

TABLE 50

	No. of hours worked	Average hourly earnings
Full-time women (manual)	39·4	111·2 pence
Full-time men (manual)	45·7	156·5 pence
Full-time women (non-manual)	36·7	143·8 pence
Full-time men (non-manual)	38·7	227·2 pence

Note: Figures include overtime payments where applicable.

into account the labour costs of mining, transportation, etc. The Coal Research Centre at Cheltenham had an employment number totalling 450 with 125 graduates as of 1976.[49] Assuming a figure of 500 for 1978, the man-hours per year utilised in such an establishment can be estimated as

$$500 \times 40 \times 52 = 1 \cdot 04 \times 10^6 \text{ man-hours per year}$$

This is based on a 40 hour week with 52 weeks in a year. To obtain the labour input to each product the D_1 factor is introduced and the value of X_1 per tonne of each product is given by

TABLE 51
$X_1 P_1$ **Values for Various Coal Processes**

Process	Products annual outputs (tonnes)	Products labour costs $X_1 P_1$ (£ per tonne)
H-coal (Case 1)	$4 \cdot 69 \times 10^6$	0·39
H-coal (Case 2)	$4 \cdot 45 \times 10^6$	0·41
COED (Case 1)	$1 \cdot 29 \times 10^6$	0·60
COED (Case 2)	$6 \cdot 67 \times 10^5$	1·23
CSF (Case 1)	$2 \cdot 52 \times 10^6$	0·74
CSF (Case 2)	$2 \cdot 50 \times 10^6$	0·75
Fischer–Tropsch (motor spirit)	$0 \cdot 94 \times 10^6$	1·02
Fischer–Tropsch (methanol)	$3 \cdot 32 \times 10^6$	0·41
NCB, hydrogenative extraction	$3 \cdot 96 \times 10^5$	3·68
NCB, supercritical extraction	$1 \cdot 0 \times 10^6$	0·80

$$X_1 = \frac{D_1 \times 1 \cdot 04 \times 10^6}{\text{product annual output}} \text{ man-hours per product tonne}$$

and

$$X_1 P_1 = £ \frac{D_1 \times 1 \cdot 85 \times 1 \cdot 04 \times 10^6}{\text{product output annually}} \text{per tonne}$$

In lieu of other labour costs information these values will be used except where specific values are quoted for specific processes. In Appendix C3 the various values of D_1 are calculated, and values of $X_1 P_1$ are shown in Table 51.

9.5 COSTS OF PRODUCING 1 TONNE OF SYNTHETIC OIL FROM COAL

The equation

$$C = X_1 P_1 + X_f P_f'(f) + X_c P_c \tag{1a}$$

is used to estimate costs of producing synthetic oils from coal using various processes. Table 48 gives the costs values estimated, while Fig. 24 shows the variation of costs with prices of coal feedstock. In summary, the mean cost of producing synthetic crude by extractive hydrogenation (or hydrogenative extraction) of coal is estimated at £ 55.00 per tonne at a coal price of £ (1978) 22·60 per tonne. In comparison, the pyrolysis process followed by char gasification will produce oils valued at higher manufacturing costs.

9.6 COSTS OF PRODUCING GASES FROM COAL

Merrick,[45] in his analysis of coal conversion processes, presents curves of costs of SNG and other fuels as functions of coal prices. From these it could be deduced that, at a coal price of £ (1978) 22·60 per tonne, the standard Lurgi process can produce SNG at a cost of £ 2·10 per GJ of energy equivalent. This approximates to about £ 2·20 per 1000 scf of pipeline SNG, or 22 pence per therm of gas. Also the advanced gasification combined cycle process can produce SNG at £ 2·0 per 1000 scf of gas or 20 pence per therm[45]. These figures seem to compare well with present average prices of domestic gases of about 18 pence per therm (£ 1·80 per

1000 scf). The details of these processes are not presented here. However, Shearer's design[17] on coal pyrolysis, followed by gasification of resulting char, will be used to estimate the costs of producing pipeline gas from coal. Other sources of gases that will be considered later are cellulose digestion and pyrolysis.

In Shearer's design 27 275 barrels of syncrude (CV = 6120 MJ per barrel), 7·08 million cubic metres of pipeline gas (CV = 34·31 MJ per cubic metre) and 1900 barrels of light hydrocarbons (CV = 4120 MJ per barrel) are the major products of the process. In the usual way of apportioning total energy requirements of the process to products it was found that 57·85% of the plant's GER is attributable to gases production. Therefore

$$D_c \text{ value} = 0·5785$$

The cost of producing these gases is given by

$$C = X_c P_c + X_1 P_1 + X_f P_f'(f) \tag{1a}$$

where all definitions remain the same. In Appendix C4 the $X_c P_c$ and $X_f P_f'(f)$ values are calculated. Also in calculations similar to those presented in Appendix C3 the value of $X_1 P_1$ is calculated. The X_f value of the process is still 0·812 as calculated in Part I. The price of coal is taken as £(1978) 22·60 per tonne and the (f) value is

$$\frac{1000 \times 1·05 \times 10^6}{26\,040 \times 10^6} = 0·04 \text{ tonne per 1000 scf of gas}$$

where $26\,040 \times 10^6$ is the calorific value of the coal used in the pyrolysis process, 29 410 tonnes per day, 57·85% is the fraction of the total attributable to gases production, and $1·05 \times 10^9$ is the calorific value of 1000 scf of product gas (1 m^3 = 35·25 ft^3).

From these the summary of costs is

$$X_c P_c = £0·38 \text{ per 1000 scf of gases}$$
$$X_1 P_1 = £0·014 \text{ per 1000 scf of gases}$$
$$X_f P_f'(f) = 0·812 \times 22·60 \times 0·068$$
$$= £1·65 \text{ per 1000 scf of gases}$$

Therefore the total cost of producing gases using the pyrolysis – gasification method is

$$C = 0.38 + 0.014 + 1.65$$
$$= £2.00 \text{ per } 1000 \text{ scf of gases}$$
$$\text{or } 20 \text{ pence per therm}$$

This value also compares well with the present price of gas. It can be seen that the energy cost factor contributes about 74% of the total costs, and proves that the process of producing gases from coal is energy intensive. Though the advanced combined process has not been discussed here we predict that the process is also energy intensive.

CHAPTER 10

Estimated Costs of Producing Syncrude from Tar Sands and Oil Shale Rocks

10.1 PREAMBLE

The tar sands and oil shale rocks sources of syncrude will be treated together because of their similarities in operation—the oil which is embedded in some form of rock is extracted by application of heat. From Part I it will be seen that the tar sand processes are slightly more energy intensive than the oil shale processes. Respective NER values per tonne of syncrude are: tar sands 16 450 MJ per tonne of syncrude, using sands of grade 13·26% by weight of bitumen; oil shale 9760 MJ per tonne of syncrude, using shale rocks grade of 20 US gallons of oil per tonne of shale rock, and a retort efficiency of 80% Fischer assay (see p. 38). The procedure for determining the costs of producing oil from these sources will be the same and the cost equation is

$$C = X_f P_f + X_l P_l + X_c P_c \qquad (1)$$

However, certain assumptions and approximations are made here. First, there are no known prices of raw shale rocks and tar sands, and so the use of prices of raw material, as in coal processes, will be difficult. Secondly, the X_f values of the two processes are tar sands 0·44 and oil shale 0·23; thus, both processes have values less than 0·5. Thirdly, Hemming[13,22,23,50] assumes that c. 80% of the total energy inputs to the processes are derived internally. These three conditions strongly favour the possibility of an approximation, namely $C = P_f$. Thus the cost equation becomes

$$C = X_f C + X_l P_l + X_c P_c$$

or

$$C = \frac{X_1 P_1 + X_c P_c}{1 - X_f} \tag{11}$$

Thus the only cost factors to be considered are labour costs, $X_1 P_1$, and capital costs, $X_c P_c$.

10.2 UNIT CAPITAL COSTS FOR PRODUCING SYNCRUDE FROM OIL SHALE

Donnel[51] has estimated the capital investment of an industry producing oil from shale rocks at a rate of 50 000 barrels per day as approximately $(1975) 1 billion. Let us suppose this amount is to be paid out in 10 years with a plant life period of 20 years. The corresponding annual charges rate will be 10%. Suppose another 5% of total investment, i.e. $P_c = 0.15$. The yearly charges will then be given as

$$X_c P_c = 0.15 \, (CI)$$

where CI is the capital investment. Note that the D_c value is 1, since oil is assumed to be the main product. Therefore

$$X_c P_c \text{ of the process} = £ (1978) \frac{1.0 \times 10^9 \times (1.1)^3}{1.95} \times 0.15$$
$$= £ 1.024 \times 10^8 \text{ per annum}$$

Annual output of syncrude from the process is 2.35×10^6 tonnes. Therefore

$$X_c P_c \text{ per tonne of product} = \frac{1.024 \times 10^8}{2.35 \times 10^6}$$
$$= £ 43.6$$

Prien[52] estimates that the capital cost of a plant producing 100 000 barrels per day is about $(1974) 720 million. As in the case of Donnel's estimate, the grade of shale rocks used in this process is about 25 gallons of oil per tonne of shale rock. With this figure of $720 million, the calculated $X_c P_c$ is £(1978) 17.2 per tonne of product oil. Hemming[23] calculated capital requirements for grades of shale rocks ranging from 5 to 50 gallons of oil per tonne of rock. From these calculations one could deduce an average value of £(1978) 15.20 per tonne of product oil. Thus, because of the variations of costs with various workers an average value will be assumed here. It is also to be noted that capital costs will vary with grades of shale rocks, but an average value of £ 20·00 per tonne of oil product is assumed here.

10.3 UNIT LABOUR COSTS FOR PRODUCING SYNCRUDE FROM OIL SHALE

Hemming[23] has considered the labour requirements of syncrude production from tar sands in two major respects. In the first instance he calculates the labour input to a process involving mining, crushing, transportation of crushed rocks over 15 km to the retort plant and disposal of spent shale 1 km from the retort plant. In the second instance he considers a process with conditions similar to those in the first case but with the disposal site 15 km away from the retort plant. The results are given in Tables 52 and 53. The figures of Table 53 are used to represent X_1 values since they are higher and it will be more prudent to use higher values of X_1 here. It should be noted that Cases 1–5 refer to retort efficiencies as discussed in Part I.

In 1974 the average labour cost in industry in the US was $ 5·20 per man-hour worked;[23] at 1978 price levels this is approximately £ 4·23 per man-hour worked. Thus the X_1P_1 values can be calculated for various grades of oil shale.

TABLE 52
Labour Requirements for the Production of 1 tonne of Syncrude
(Retort to Disposal Distance = 1 km)

Shale grade (gallons of oil per tonne of rock)	X_1 values (man-hours per tonne of syncrude)				
	Case 1	Case 2	Case 3	Case 4	Case 5
5	–	96·99	31·79	19·41	15·80
10	19·48	11·34	9·49	8·21	7·59
15	8·30	6·61	6·05	5·56	5·33
20	5·61	4·89	4·61	4·38	4·25
30	4·01	3·71	3·58	3·46	3·39
40	3·21	3·05	2·98	2·91	2·87
50	2·72	2·63	2·58	2·54	2·52

(Source: ref. 23.)

TABLE 53
Labour Requirements for the Production of 1 tonne of Syncrude
(Retort to Disposal Distance = 15km)

Shale grade (gallons of oil per tonne of rock)	X_l values (man-hours per tonne of syncrude)				
	Case 1	Case 2	Case 3	Case 4	Case 5
5	–	100·9	33·02	20·15	16·38
10	10·20	11·73	9·82	8·48	7·85
15	8·57	6·82	6·23	5·73	5·49
20	5·77	5·03	4·74	4·50	4·37
30	4·11	3·80	3·66	3·54	3·47
40	3·27	3·11	3·03	2·97	2·93
50	2·76	2·67	2·63	2·58	2·56

(Source: ref. 23.)

TABLE 54
Cost of Producing 1 tonne of Syncrude from Oil Shale.
Fischer assay = 95%, $P_c = 0·15$, $P_1 = £4·23$ per man-hour

Shale grade (gallons of oil per tonne of shale)	Labour requirements, $X_l{}^a$ (man-hours per tonne)	Average capital cost, $X_c P_c$ (£)	Energy factor, X_f	Cost of syncrude (£ per tonne)
5	33·02	20·00	1·531	–
10	9·82	20·00	0·347	94·20
15	6·23	20·00	0·242	61·20
20	4·74	20·00	0·202	50·20
30	3·66	20·00	0·174	43·00
40	3·03	20·00	0·158	39·00
50	2·63	20·00	0·148	36·50

[a] Obtained from ref. 23.

10.4 COST OF PRODUCING 1 TONNE OF SYNCRUDE FROM OIL SHALE

$$P_c = 0·15$$
$$X_c P_c = £20 \text{ per tonne (on average)}$$
$$P_1 = £4·23 \text{ per man-hour worked}$$
$$X_f = \text{values already calculated}$$
$$C = \frac{X_c P_c + X_l P_1}{1 - X_f}$$

The values of C are given in Tables 54–57 for various grades and retort efficiencies. For a specific example it will cost £ 50·20 to produce 1·0 tonne of syncrude from oil shale rock of grade 20 gallons of oil per tonne of rock applying a retort efficiency of 95% Fischer assay. The same shale grade treated to 100% retort efficiency will produce oil at a cost of £ 50·00 per tonne, while with a retort efficiency of 90% the cost will be £ 52·10 per

TABLE 55
Cost of Producing 1 tonne of Syncrude from Oil Shale.
Fischer assay = 100%, $P_c = 0·15$, $P_l = £ 4·23$ per man-hour

Shale grade (gallons of oil per tonne of shale)	Labour requirements, $X_l{}^a$ (man-hours per tonne)	Average capital cost, $X_c P_c$ (£)	Energy factor, X_f	Cost of syncrude (£ per tonne)
5	20·15	20·00	0·737	400·00
10	8·48	20·00	0·307	80·60
15	5·73	20·00	0·229	57·40
20	4·50	20·00	0·220	50·00
30	3·54	20·00	0·171	42·20
40	2·97	20·00	0·156	38·60
50	2·58	20·00	0·147	36·20

[a] Obtained from ref. 23.

TABLE 56
Cost of Producing 1 tonne of Syncrude from Oil Shale.
Fischer assay = 100%, $P_c = 0·15$, $P_l = £ 4·23$ per man-hour

Shale grade (gallons of oil per tonne of shale)	Labour requirements, $X_l{}^a$ (man-hours per tonne)	Average capital cost, $X_c P_c$ (£)	Energy factor, X_f	Cost of syncrude (£ per tonne)
5	16·38	20·00	0·578	211·60
10	7·85	20·00	0·288	74·70
15	5·49	20·00	0·222	55·60
20	4·37	20·00	0·192	47·60
30	3·47	20·00	0·169	41·70
40	2·93	20·00	0·155	38·30
50	2·56	20·00	0·146	36·10

[a] Obtained from ref. 23.

TABLE 57
Cost of Producing 1 tonne of Syncrude from Oil Shale.
Fischer assay $= 90\%$, $P_c = 0.15$, $P_1 = £4.23$ per man-hour

Shale grade (gallons of oil per tonne of shale)	Labour requirements, X_l^a (man-hours per tonne)	Average capital cost, $X_c P_c$ (£)	Energy factor, X_f	Cost of syncrude (£ per tonne)
5	100·90	20·00	–	–
10	11·73	20·00	0·409	117·80
15	6·82	20·00	0·259	66·00
20	5·03	20·00	0·208	52·10
30	3·80	20·00	0·177	43·80
40	3·11	20·00	0·160	39·50
50	2·67	20·00	0·149	36·80

^{a} Obtained from ref. 23.

tonne. Similar comparisons can be drawn from the tables given above. Figure 25 gives the graphical presentation of these costs. The 20 gallons per tonne of rock seems to be a cut-off grade. Below this grade costs vary remarkably, whilst above this grade costs seem to be fairly insensitive to variations in grades of shale rocks.

10.5 UNIT CAPITAL COSTS FOR PRODUCING SYNCRUDE FROM TAR SANDS

Swabb[6] quotes the initial investment at the GCOS as $ (1967) 253 million for a plant with an annual capacity of 2·2 million tonnes of syncrude. Hemming[50] indicates that capital investment at the GCOS was $ (1964) 171 million for the same output. Using the figures of Hemming and an annual charges rate of 15% the capital costs per tonne of product, $X_c P_c$, are calculated in Appendix C5, and are approximately £ (1978) 22·30 per tonne of syncrude.

10.6 UNIT LABOUR COSTS FOR PRODUCING SYNCRUDE FROM TAR SANDS

Canada Yearbook 1974[54] gives the average industrial wages for non-office labour force as

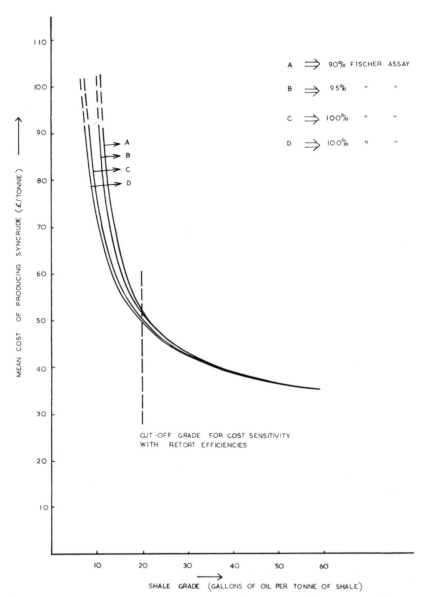

Fig. 25.   *The cost of syncrude production as a function of oil shale grade and retort efficiency.*

(i) construction—$ (1972) 6·00 per man-hour worked,
(ii) manufacturing and other industries—$ (1972) 4·10 per man-hour worked.

Using an average value for all industries of $ (1972) 5·0 per man-hour, the cost of labour, P_1, will be

$$\$ (1972) \; 5\cdot0 \text{ per man-hour}$$

or

$$\pounds (1978) \; 4\cdot40 \text{ per man-hour worked*}$$

Oliver[55] gives an indication of the labour force required for tar sands processing into synthetic crude. Approximately 750 engineers will be required by developers and contractors to complete the engineering job that has a lead time of about 2–3 years. Also peak skilled construction labour will be required at an estimated number of 6000 men for each project with a total labour requirement of up to 20 000 man-years over the 5–6 years of construction work. Then about 2200 skilled men will be required for constant operation and maintenance. Assuming the project life is 25 years and amortising the labour requirements for construction and engineering over this period, the estimated labour requirements for the entire process will be given as:

engineering	=	45 man-years
construction	=	800 man-years
operation and maintenance	=	2200 man-years

Total   3045 man-years

With a 40-hour working week, this labour requirement approximates

$$3045 \times 40 \times 52 = 6\cdot33 \times 10^6 \text{ man-hours per annum}$$

Therefore P_1X_1 for the process of tar sands to oil may be estimated as

$$\frac{4\cdot4 \times 6\cdot33 \times 10^6}{2\cdot2 \times 10^6} = \pounds\, 12\cdot1 \text{ per tonne of syncrude}$$

where $2\cdot2 \times 10^6$ is the annual tonnage output of the plant.

* $\pounds (1978) \; 1\cdot0 = \$ (1978) \; 1\cdot95$; $\$ (1978) = \$ (1972) \times (1\cdot1)^6$.

10.7 COST OF PRODUCING 1 TONNE OF SYNCRUDE FROM TAR SANDS

The cost equation is

$$C = \frac{X_1 P_1 + X_c P_c}{1 - X_f}$$

and $X_1 P_1$ and $X_c P_c$ have been calculated. The X_f value is calculated in Part I and is equal to 0·388 for a plant treating sand grade of 13·26% by weight of bitumen.

Therefore for the GCOS typical grade of 13·26%

$$C = \frac{22·3 + 12·1}{1 - 0·388}$$

$$= £ 56·20 \text{ per tonne of syncrude}$$

In Table 58 the variation of costs of production with grades of tar sands are presented, and Fig. 26 gives a graphical presentation. An economical

TABLE 58
Variation of Costs of Syncrude Production with Grade of Tar Sands.
($P_1 = £4·40$ per man-hour, average capital costs $(X_c P_c)^a$ of tar sands process = £ 22·30 per tonne of product.)

Tar sand grade	Labour requirements, X_1^b (man-hours per tonne)	Energy factor, X_f	Cost of syncrude product (£ per tonne)
5·0	8·93	1·10	–
6·0	7·02	0·815	287·50
7·0	5·78	0·67	143·30
8·0	4·92	0·577	104·00
9·0	4·28	0·517	85·20
10·0	3·78	0·473	74·00
11·0	3·39	0·441	66·60
12·0	3·07	0·415	61·20
13·26	2·75	0·388	56·20
15·0	2·40	0·363	51·60
18·0	1·97	0·332	46·40
20·0	1·76	0·318	44·10

[a] In lieu of published variations of investments with tar sand grades, the average value of £ 22·30 per tonne of product will be used. This is bound to vary with grades in practice.
[b] X_1 value = (ratio of weight of grade to that of 13·26%) × 12·1/4·4.

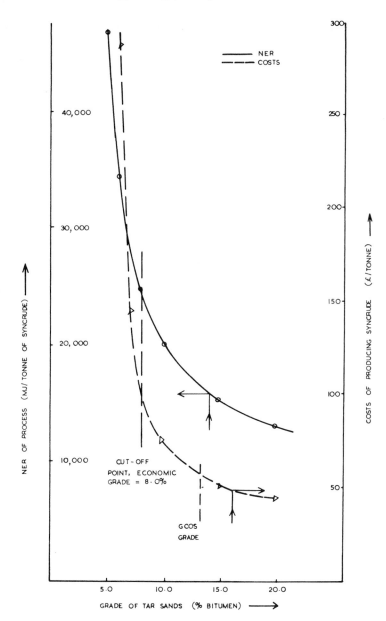

Fig. 26. *Variation of costs of syncrude and NER of process with grades of tar sand.*

cut-off grade is *c*. 8·0% of bitumen by weight, and at this grade a production cost of about £150 per tonne of oil is anticipated.

Oliver[55] gives the estimated cost of syncrude from tar sands, at 1976 dollar levels, as $14·20 per barrel; at 1978 levels this is approximately £63·00 per tonne of syncrude. Thus, our calculated value of £56·20 per tonne at the GCOS 13·26% grade seems to compare well with Oliver's estimates.

CHAPTER 11

Estimated Costs of Producing Syncrude from Organic Matter

11.1 PREAMBLE

Trying to cost cellulose-based processes is potentially very difficult. In some communities the cost of refuse or crop wastes is usually viewed as the entire cost of transportation and collection. Some disposal units quote costs of disposing of certain quantities of refuse or wastes as the values of these wastes. In trying to cost the conversion of municipal wastes (say) to oil or gases, one has many parameters to consider: the labour and capital costs of disposal techniques; the value of land for tipping of this refuse; and some other factors such as incentive bonuses given to the population of the municipality for segregation of wastes in combustible and non-combustibles. However, in this work the assumptions that will be made are that land usage will be government financed, i.e. all lands will be government-owned and hence land costs will not be considered. Also all labour costs in collection and transportation aspects of the process will be treated purely as labour costs. Capital costs for collection vehicles, or tractors (where applicable), conversion plants, fertilisers, chemicals and other utilities will be treated under capital investments and subjected to the same annual charges rate. The energy factor will include those of conversion plant, transport vehicles and collection equipment. Also, since more than 80% of the total energy input to the process will be derived on-site, the assumption $C = P_f$ in the cost equation will be made. That is to say that the cost equation in all cellulose-based processes, except the pyrolysis of wastes, will be given by

$$C = \frac{X_c P_c + X_1 P_1}{1 - X_f}$$

Also, all other sociological factors such as taxation, grants, etc., will be neglected—again costs estimated will serve as 'guide-costs'.

11.2 UNIT CAPITAL COSTS OF PLANT PRODUCING OIL FROM MUNICIPAL REFUSE

In Part I the energy requirements of such a plant were determined. In this section the costs will be estimated using the energy factor, X_f.

Kaufman and Weiss[25] estimate that the capital cost of a plant treating 36 tonnes of municipal wastes per day and producing about 13 tonnes of oil daily will be about $ (1975) 1 116 906; they further estimate that this money should be paid out over 10 years. However, since this process is relatively new, compared to coal processes, such a plant will be given a 20-year lifetime here and a payout period of 7 years. Also 7% of capital investment will be charged for maintenance services, giving an annual capital charges rate of approximately 17%. With this the annual capital charges of the process will be given by

$$X_c P_c = R'(\text{CI})D$$

where $R' = 17\%$;
            $\text{CI} = $ capital investment;
            $D = $ fraction of product in all process outputs
            $= 0.83$ (Part I).

Thus

$$X_c P_c = 0.17 \times 782\ 422^* \times 0.83$$
$$= £\ 110\ 400 \text{ per annum}$$

But the process produces 12·95 tonnes daily for a total period of 260 days per year. Therefore the annual output = 3367 tonnes. Hence

$$X_c P_c = \frac{110\ 400}{3367}$$
$$= £\ 32.80 \text{ per tonne of oil}$$

$^*£(1978)1.0 = $(1978)\ 1.95;\quad $(1978) = $(1975) \times (1.1)^3.$

11.3 UNIT LABOUR COSTS OF PLANT PRODUCING OIL FROM MUNICIPAL REFUSE

Labour costs will be those due to the plant staff and transportation or collection costs of the refuse. As shown in Appendix C6 transportation costs of refuse in the Greater London area are estimated to be £86·40 per day for collection of 48 tonnes. Note that 48 tonnes have to be collected in order to have 36 tonnes of combustibles (the Holden case study has refuse of constitution 25% non-combustibles). In the Greater Manchester area costs of the order of £1·60–£1·80 per tonne of refuse are quoted, i.e. approximately £82·00 for 48 tonnes. A cost of £86·40 for 48 tonnes is assumed here.

Kaufman and Weiss[25] estimate the staffing figure of such a plant as approximately 10 because of the highly automated nature of the plant. However, considering operational problems, staffing capacity will be put at 30–40 for three shifts, and it will be assumed that each person will have a 40-hour per week schedule throughout the year; earnings are assumed to be about £1·86 per man-hour worked. Hence

$$\text{labour costs} = 86 \cdot 4 + \frac{35 \times 40 \times 52 \times 1 \cdot 86}{260}$$

$$= \pounds 607 \cdot 4 \text{ per operating day}$$

But tonnage product output is 12·95 each operational day. Therefore

$$X_1 P_1 = \frac{607 \cdot 4}{12 \cdot 95} = \pounds 46 \cdot 9 \text{ per tonne of oil}$$

11.4 COST OF PRODUCING 1 TONNE OF OIL FROM MUNICIPAL REFUSE

For a plant of capacity of 36 tonnes per day the cost of producing 1 tonne of oil will be

$$C = \frac{46 \cdot 9 + 32 \cdot 8}{1 - X_f}$$

In Part I the X_f value was calculated as $X_f = 0 \cdot 41$. Therefore

$$C = \frac{79 \cdot 7}{0 \cdot 59}$$

$$= \pounds 135 \cdot 00 \text{ per tonne of oil}$$
$$\text{or } \pounds 19 \cdot 30 \text{ per barrel of oil}$$

TABLE 59
Variation of Cost Factors of Refuse Processing Plants with Plant Capacities

Plant refuse capacity (tonnes per day)	Plant capital costs ($ million)	Refuse transportation costs (£ per day)	Plant labour costs (£ per day)
36	1·487	86·4	521
100	2·742	240	521
500	7·214	1 200	521
1 000	10·914	2 400	521
2 000	16·504	4 800	521

TABLE 60
Cost of Producing Oil from Municipal Refuse as a Function of Plant Capacity

Plant capacity (tonnes per day)	Energy factor, X_f	Capital costs $(X_c P_c)$ per tonne of product (£)	Plant labour costs per tonne of product (£)	Transportation costs per tonne of product (£)	Total cost of producing 1 tonne of product oil (£)
36	0·410	32·80	40·40	6·67	135·00
100	0·382	21·80	14·50	6·67	70·00
500	0·378	11·50	2·90	6·67	34·00
1000	0·375	8·60	1·50	6·67	27·00
2000	0·370	6·70	0·75	6·67	22·50

£ 19·30 per barrel of oil is almost competitive with present day price levels. Moreover, there is no question that the capacity of the plant can be increased. Actually Kaufman and Weiss carried out designs for capacities of plant up to 2000 tonnes per day of combustible refuse. In Appendix C7 the various labour and capital costs for plants with capacities greater than 36 tonnes per day are calculated. Tables 59 and 60 present a summary of the various costs. It is seen that the cost of producing oil with a plant of capacity of 500 tonnes per day is *c.* £ 34 per tonne of product, while with plants of capacities 1000 and 2000 tonnes per day costs of £ 27·00 per tonne and £ 22·50 per tonne of product are obtained, respectively. Between capacities of 100 tonnes per day and 500 tonnes per day of refuse, there is a wide cost difference (£ 70·00 and £ 34·00 per product tonne, respectively). Figure 27 shows the variation of cost with plant refuse capacity. For values

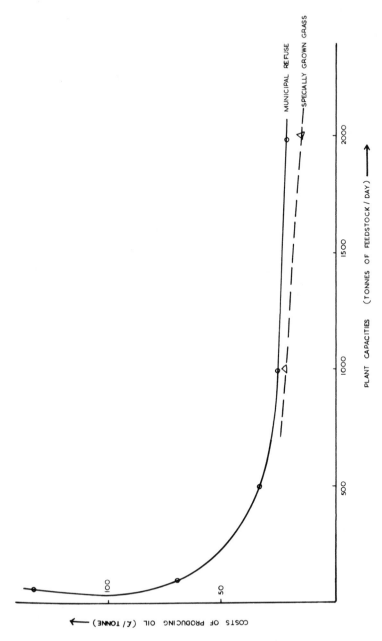

Fig. 27. Cost of producing oil from refuse and grown crops as a function of plant capacity.

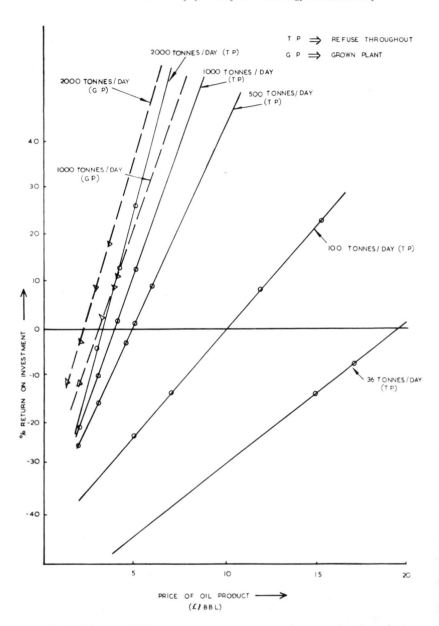

Fig. 28. Cellulose to oil plant returns on investments as a function of product oil prices and plant capacities.

above 500 tonnes capacity, costs seem to vary very gradually, and thus one can comfortably say that the economic cut-off capacity in this process is 500 tonnes per day. Figure 28 gives a graphical presentation of percentage returns on investments of the various plant capacities, from which it will be seen that capacities of 500 + tonnes per day give very attractive returns on investments, while capacities of 100 and 36 tonnes per day give low percentage returns. As an example, at on oil price of £ 10·0 per barrel the 36 tonnes plant does not break even while the 100 tonnes plant barely breaks even and plants of capacities above 500 tonnes per day give very large percentage returns. For comparison purposes the returns of plants based on specially grown crops are also shown in Fig. 28. The returns are calculated to show the viability and attractiveness of cellulose-based processes. This aspect is discussed in detail later.

11.5 UNIT CAPITAL COSTS OF PLANT PRODUCING OIL FROM SPECIALLY GROWN CROPS

In the case of crops being grown specially for conversion into oil, e.g. grass conversion to oil, additional investments are required for the purchase of transport vehicles and field tractors and equipment. Plant operations will be assumed to be similar to those of refuse-based plants, i.e. 260 operating days in a year. The transportation modes and tractor requirements have been discussed in Part I. It was noted that a plant needing 1000 tonnes of hay per day will require about 11 trucks for the collection of these feedstocks. Also, it was calculated that about 185 tractors will be required over a period of 6·67 years to cultivate 13 200 hectares of land. Suppose the cost of each truck is £4000, and each tractor costs £5500, then the additional initial investment required for this process will be

$$(11 \times 4000) + (5500 \times 185) = £1·06 \text{ million}$$

The capital investment for plant of capacity 1000 tonnes per day was calculated as £5·74 million (refuse-based plant). Therefore the total plant investment initially, in this case, is £6·80 million. Using the same procedures of costing, i.e. D value $= 0·83$, annual charges rate $= 17\%$, the capital costs of output product will be

$$X_c P_c = \frac{0·17 \times 0·83 \times 6·80 \times 10^6}{360 \times 260}$$
$$= £10·35 \text{ per tonne}$$

Note: in the energy requirements calculations it was noted that 1000 tonnes of refuse should give 360 tonnes of oil.

11.6 UNIT LABOUR COSTS OF PLANT PRODUCING OIL FROM SPECIALLY GROWN CROPS

In Part I it was noted that a standard tractor has an average yearly performance of 900 hours. Thus for 185 tractors a total of 185×900 or 166 500 hours operation is possible per year. Field workers driving these tractors will therefore have wages equal to

$$£(166\ 500 \times \text{hourly rate}) \text{ per year}$$

From the amendments of 1978 to the Agricultural Workers' Pay Act[56] the average hourly pay rate of agricultural workers is calculated to be £1·30. Hence the tractor drivers will be paid

$$166\ 500 \times 1·30 = £216\ 450 \text{ per annum}$$

giving a daily labour cost of

$$\frac{216\ 450}{260} = £832·50$$

Plant operators' wages are based on 35 persons receiving £1·86 per hour for 40 hours a week throughout the year, giving a labour cost of

$$\frac{35 \times 40 \times 52 \times 1·86}{260} = £521·0 \text{ per operational day}$$

Truck drivers are assumed to receive wages similar to the tractor drivers giving another labour cost of

$$11 \times 12 \times 1·30 = £172 \text{ per day}$$

where each truck operates 12 hours per day and needs two drivers (6 hours per day each).

Thús a total labour cost of £1526·0 will be incurred each operational day. For a product output of 360 tonnes per day, the labour costs are given by

$$X_1 P_1 = \frac{1526·0 \times 0·83}{360}$$
$$= £3·52 \text{ per tonne of product}$$

11.7 COST OF PRODUCING 1 TONNE OF OIL FROM SPECIALLY GROWN CROPS

For a 1000 tonnes capacity plant the cost, C, is given by

$$C = \frac{3 \cdot 52 + 10 \cdot 35}{1 - X_f}$$

From Part I, the X_f value for specially grown crops for a 1000 tonnes per day plant is 0·287. Therefore

$$C = \frac{3 \cdot 52 + 10 \cdot 35}{0 \cdot 713}$$
$$= £\,20 \cdot 20 \text{ per tonne of oil}$$

Similar calculations were done for a 2000 tonnes per day plant and the following results were obtained:

$$X_f = 0 \cdot 280$$

giving

$$C = £\,15 \cdot 10 \text{ per tonne of oil}$$

The percentage returns on investment of these two capacities are presented in Fig. 28. Comparative discussions are made on these later.

11.8 UNIT CAPITAL COSTS OF PLANT PRODUCING METHANE GAS FROM CROP MATTER

Since the product is gas the cost estimate of this process will be based on 1000 scf of gas and 1 therm (100 scf) of gas. In the energy requirements section (Part I) two cases of plant matter were considered. The first used leafy material such as grasses and the second used woody material. In the two cases the energy factors were calculated as leafy matter 0·28 and woody plant or crop 0·336. Under the leafy matter consideration, two sub-cases were considered—the question of using crop wastes, resulting from harvesting and planting periods as feedstock or the use of specially grown crops as feedstock. However, since in practice there may be no distinction of the matter used, an average value of the two estimated costs will be taken.

Sitton and Gaddy[29] give the cost of a plant processing hay, etc., to methane gas as $ (1974) 45·7 million including some 30% contingency costs. Bringing this to 1978 price levels the plant investment is about £ (1978) 34·3 million. To this must be added the additional initial capital cost of trucks, tractors, etc.

11.8.1 Case for crop wastes

In Part I the additional investment for purchase of trucks only was estimated at £ 0·192 million. The annual charges rate will still be put at 17%. The value of D is taken here as unity because all investments are geared towards producing methane. Note that in the case of energy apportionment the D value was taken as 0·94. This is because in energy requirement considerations it was taken that some form of energy was used to produce the byproduct CO_2. Hence

$$X_c P_c = 0·17 (34·3 + 0·192) \times 10^6$$
$$= £ 5·864 \times 10^6 \text{ per annum}$$

Since the output of such a plant is given as 50 million scf of methane per day[29]

$$X_c P_c \text{ per daily 1000 scf of gas} = \frac{5·864 \times 10^6}{50 \times 10^3 \times 330}$$

where operation is carried out for 330 days per year, i.e.

$$X_c P_c = £ 0·36 \text{ per 1000 scf of gas}$$

11.8.2 Case for grown crops

Additional investments for field tractors and equipment were calculated to be £ 4·46 million. The lorries and conversion plant investments remain the same. Hence $X_c P_c$ in this case is given by

$$X_c P_c \text{ per 1000 scf} = 0·36 + \frac{0·17 \times 4·46 \times 10^6}{50 \times 10^6 \times 330}$$
$$= £ 0·406$$

Thus a mean value for these two cases is

$$\frac{0·36 + 0·406}{2} = £ 0·383 \text{ per 1000 scf of gas}$$

11.8.3 Capital costs for woody matter as feedstock

In this case only grown crops are considered since most of the woody plants will have to be grown and collected. Fraser[42] gave the plant investment for an optimum plant having 16 pretreatment – digestion trains (see Part I) as $ (1974) 250 million. At 1978 price levels this is approximately £ (1978) 187 million. Additional investments for trucks and tractors are put at £ 12·32 million. Therefore

$$X_c P_c = 0·17(12·32 + 187) \times 10^6$$
$$= £ 34 \times 10^6 \text{ per annum}$$

Process output is given as 92·4 million scf of gas per day or 30 492 million scf per year of 330 operating days. Therefore

$$X_c P_c \text{ per daily } 1000 \text{ scf} = \frac{34 \times 10^6}{30\,492 \times 10^3}$$
$$= £ 1·12$$

It can be seen that the capital costs for woody material are greater than those for leafy feedstock. This is quite reasonable, since in the former case many more pretreatment – digestion reactors are needed, more CO_2 is produced and more separation stages are needed.

11.9 UNIT LABOUR COSTS OF PLANT PRODUCING METHANE GAS FROM CROP MATTER

Labour requirements in such a plant are very uncertain, as the process is slow and requires many chemical handling and separation operations. To be in line with other cellulose-based processes it will be assumed that operational staff will be 35 in number for three shifts. Wages are £ 1·86 per man-hour worked and operation will be for 330 days per year, with each operator working 40 hours per week. Thus plant labour costs will be

$$\frac{35 \times 40 \times 52 \times 1·86}{330} = £ 410 \text{ per day of operation}$$

In the case of crop wastes (Part I) it was calculated that farmers obtain £ 13 140 per day as labour wages for collecting the waste, and truck drivers will receive £ 750 per day, giving a total labour cost of £ 14 300 per day. The unit labour requirement per 1000 scf of gas will therefore be

$$\frac{14\,300}{50 \times 10^3}$$

where 50 million scf is the daily gas output, i.e.,

$$X_1 P_1 \text{ per } 1000 \text{ scf of gas} = £ 0·290$$

In the case of grown crops as feedstock to plant, labour cost requirements are: plant operators £410 per day, truck drivers £750 per day, and tractor drivers £3646 per day (Part I), giving a total of £4806 per day. Hence

$$X_1 P_1 \text{ per } 1000 \text{ scf of gas} = \frac{4806}{50 \times 10^3}$$
$$= £0 \cdot 11$$

A mean of these is given by

$$\frac{0 \cdot 11 + 0 \cdot 29}{2} = £0 \cdot 21 \text{ per } 1000 \text{ scf of gas}$$

For woody crop matter as feedstock the calculated labour costs (Part I) are £410 per day for plant operators, £9740 per day for tractor drivers and £1623 per day for truck drivers. Thus, total labour costs are £11 773 per day and with a daily output of 92·4 million scf, the $X_1 P_1$ per 1000 scf is given as

$$\frac{11\ 773}{92 \cdot 4 \times 10^3} = £0 \cdot 13$$

11.10 COST OF PRODUCING 1000 SCF OF METHANE FROM CROP MATTER

11.10.1 Leafy matter such as hay
X_f value of process $= 0 \cdot 28$
$X_1 P_1$ per 1000 scf of gas $= £0 \cdot 21$
$X_c P_c$ per 1000 scf of gas $= £0 \cdot 383$

$$C = \frac{X_1 P_1 + X_c P_c}{1 - X_f}$$
$$= \frac{0 \cdot 21 + 0 \cdot 383}{1 - 0 \cdot 28}$$
$$= £0 \cdot 824$$

This is equivalent to 8·3 pence per therm of methane gas.

11.10.2 Woody matter such as sycamore
$X_f = 0 \cdot 336$
$X_1 P_1$ per 1000 scf of gas $= £0 \cdot 13$
$X_c P_c$ per 1000 scf of gas $= £1 \cdot 12$

$$C = \frac{X_1 P_1 + X_c P_c}{1 - X_f}$$
$$= \frac{0 \cdot 13 + 1 \cdot 12}{1 - 0 \cdot 336}$$
$$= £ 1 \cdot 90$$

This is equivalent to 19·0 pence per therm of methane gas.

11.11 COST OF WASTES PYROLYSIS PRODUCTS FROM 200 TONNES PER DAY SAN DIEGO PLANT

The intriguing aspect of this type of process is that the products (namely gases, oil and char) are usually of low quality. As mentioned in Part I, cost estimates of producing such low quality products may be misleading. To be of any use these products have to be treated further, and further costs are incurred. However, for reasons given in Part I, it is necessary to carry out such cost estimates. The energy requirement factor is high ($X_f = 0 \cdot 70$) and the approximation of $C = P_f$ will not give realistic cost values. Moreover, the majority of energy inputs do not come from the refuse heat content only. Considerable heat is required to fuel the burners so this approximation is inappropriate. Hence the cost equation will be

$$C = X_c P_c + X_1 P_1 + X_f P_f'(f) \tag{1a}$$

where definitions are the same as in Chapter 9.

11.11.1 Unit capital cost of process
Levy[39] estimates the capital costs of the San Diego pyrolysis plant to be $ (1974) 6·3 million, equivalent to £ (1978) 4·7 million. Pyrolysis of wastes is a process that has been investigated a good deal and thus an annual capital charge rate of 14% will be fixed, implying a payout period of about 9 years. One other assumption that will be made is that the cost of production will be based on the thermal equivalent of the total products, i.e. the products as a whole will be taken and costs will be based on their heat values equivalent. As an example, the San Diego 200 tonnes per day plant produces 33·8 tonnes of oil per day (heat value of 24 400 J per gram), 11·1 tonnes of char per day (heat value of 21 000 J per gram) and 17 million litres of gases per day (heat value of 20 000 J per litre). The total heat value of the products is $1·40 \times 10^6$ MJ per day or 1330 million Btu per day. The capital costs will be based on million Btu heat value per day.

$$X_c P_c = 0.14(4.7 \times 10^6) \times D_c, \text{ (where } D_c = 1)$$
$$= £ 660\ 820 \text{ per annum}$$

Therefore

$$X_c P_c \text{ per million Btu equivalent of product} = \frac{660\ 820}{1330 \times 330}$$
$$= £ 1.51$$

where the plant is operated for 330 days per year.

11.11.2 Unit labour cost of process

Levy[39] estimates a labour cost requirement of \$ (1974) 360 000 annually for the San Diego plant, equivalent to about £ (1978) 269 720. Hence $X_l P_l$ per million Btu equivalent is

$$\frac{269\ 720}{1330 \times 330} = £ 0.614$$

11.11.3 Unit energy cost of process

The energy cost factor of the process, X_f, is 0.70, and the contribution to total costs by energy is given by $X_f P_f'(f)$, where P_f' is the cost of collecting and transporting refuse to the plant site and (f) is the ratio of tonnes of refuse to 1 million Btu equivalent of products.

It has been stated that the total GER of the process (Part I) is 2.37×10^6 MJ per day and the calorific value of input refuse is 9.7×10^3 MJ per tonne, i.e. 244 tonnes equivalent of refuse is consumed by the process each day. The product output each day is 1330 million Btu; thus

$$(f) = \frac{244}{1330}$$
$$= 0.183 \text{ tonnes per million Btu}$$

The cost of transporting the refuse to the pyrolysis plant will be assumed to be the same as in Section 11.3, i.e. £ 1.80 per tonne of refuse. Therefore

$$X_f P_f'(f) = 0.70 \times 1.80 \times 0.183$$
$$= £ 0.23 \text{ per million Btu of product}$$

11.11.4 Mean cost of producing gases, oil and char by pyrolysis of wastes

$$C = X_c P_c + X_l P_l + X_f P_f'(f)$$
$$= 1.510 + 0.614 + 0.23$$
$$= £ 2.35 \text{ per million Btu of product}$$

TABLE 61
Costs^a of Producing Synthetic Fuels from Wastes by the Pyrolysis Process

Cost factors considered	Plant capacities (tonnes of refuse per day)		
	200	1000	2000
Plant investments in £(1978) million	4·7	12·34^b	18·71
Capital cost requirements, X_cP_c, per million Btu of product (£)	1·51	0·79	0·60
Labour cost requirements, X_1P_1, per million Btu of product (£)	0·62	0·13	0·06
Energy cost requirements, X_fP_f' (f), per million Btu of product (£)	0·23	0·22	0·21
Synthetic fuels (gas, oil and char) production (million Btu equivalent per day)	1 330	6 650	13 300
Synthetic fuels production (tonnes per day)	72·3	362	723
Estimated cost of producing 1 million Btu equivalent of synthetic fuels (£)	2·36	1·14	0·87

^{a} Calculations based on the San Diego Garret Process Plant.
^{b} Estimated using the six-tenths rule.

Note: in terms of natural gas equivalent, 1 million Btu is equal to 1000 scf. Thus in essence the cost value above indicates that if municipal waste is pyrolysed and the products utilised to generate heat locally the lowest price for which these products could be sold, for the process to just break even, is £2·40 for every 1000 scf of natural gas equivalent of product utilised. In comparison to other cellulose-based processes this is not quite competitive. Biological conversion of cellulose to methane should give products which can sell for as low as £0·82 per 1000 scf of gas using grass, and as low as

£ 2·0 per 1000 scf of gas using wood. Also, liquefaction of cellulose should give products that can sell for as low as £ 0·50–0·60 per 1000 scf equivalent of methane.

However, analyses carried out for plants processing 1000 tonnes per day and 2000 tonnes per day of refuse, respectively, show that lower costs of production can be achieved. In Appendix C8, summarised in Table 61, the costs of producing gases and oils from pyrolysis of refuse in a plant of capacity greater than 200 tonnes per day are given.

CHAPTER 12

The Utilisation of Synthetic Fuels

12.1 POTENTIAL PRODUCTION OF SYNTHETICS

The production of synthetic oils and gases in many countries is under increasing investigation; some countries such as the US and Canada are in the advanced stages of these experimental trials, and South Africa has operating plants. It is expected that by 1985 synthetic oils and gases will play some significant role in the energy system of the US. With its large oil shale rocks deposits, in particular in Colorado, its large coal deposits and its large expanse of agricultural land the US rightly has high hopes for synthetics to provide much of its domestic energy in the future, and even the UK has abundant coal reserves.* Tropical countries could count on cellulose for future energy. With these statements it becomes necessary to view the approximate yields of synthetics from solid fuels.

In coal-based processes yields are typically in the region of 60% with respect to thermal efficiencies. The Fischer–Tropsch's Sasol-type process has an efficiency of $c. 41\%$. In tonnes equivalent output the liquefaction processes seem to have advantages over other processes. Moreover, operating conditions for these liquefaction conversion processes are better than those of other coal processes (see Appendices A1–A6). Thus here liquefaction processes will be used as representative coal processes and they will be used to compare with other conversion (cellulose, sand, shale) processes.

---

* A recent British Petroleum Briefing Paper, 1980, estimates that the UK has 1% of world economically recoverable hard coal reserves; the US (28%), USSR (22%) and China (21%).

In terms of synthetic oil production a typical CSF process yields about 0·30–0·40 tonne of oil per tonne of coal fed for conversion purposes. The methanol process can be considered as a unique coal process because the product is not oil. In this process about 0·35 tonne of methanol is produced per tonne of coal feed. In contrast, a tonne of shale oil is produced from 12·3 tonnes of shale rock[13] whilst a tonne of syncrude is obtained from 8·38 tonnes of tar sands.[50] The grades of shale rocks and tar sands considered in quoting the figures above are 20 US gallons of shale oil per tonne of shale rock and 13·26% by weight of bitumen in tar sands. It can be seen that yields are about 0·08 tonne of oil per tonne of shale rock for shale rock processing, and 0·12 tonne of oil per tonne of tar sands in the case of tar sands processing. With cellulose the yield is typically 2·1 barrels or 0·3 tonne of oil per tonne of cellulose processed.[25] Cellulose has the additional credit of 10 000 scf of methane gas per tonne of cellulosic material digested anaerobically.[29,42] Pyrolysis of coal followed by the resulting char gasification yields about 9000 scf of gas per tonne of coal. In straight coal gasification processes, e.g. the Lurgi process, 21 000 scf of pipeline gas is obtainable per tonne of coal gasified. In tonnes of oil equivalent, these gaseous yields are approximately 0·25 tonne of oil per tonne of cellulose, and 0·50 tonne of oil per tonne of coal gasified. These approximate weight ratios have been calculated to bring to the mind of the reader the levels of yield expected in these conversion processes. Obviously coal seems to give the highest yield of oil equivalent per tonne of raw material and the future use of synthetic fuels has to be considered in line with product yield and raw materials abundance. This aspect is treated later. However, in considering the various solid fossil fuels above it must be noted that the properties of these fuels vary and yields will also vary with these properties. For example, different classes of coal, say bituminous, sub-bituminous, lignite, etc., give different yields of oil.

12.2 USES OF SYNTHETIC FUEL PRODUCTS

12.2.1 Reserves of raw materials

Penner and Icerman[30] give projections of current and future demands and reserves of oil. A typical projection is that in 1971 the world's total reserve of crude oil was about 86 billion tonnes with an expected life supply period of 35 years. In the same year natural gas reserves were 50·5 trillion cubic metres with a life span of about 37 years. Earl[36] projects that in 1980 the world is expected to produce and consume about 1·35 billion cubic metres of fuelwood. Of these about 46% will be produced in Asia Pacific, 23% in Africa and 18% in Latin America; Europe, USSR and North America will account for about 13%. Penner and Icerman[30] give the world total coal reserves, as of 1974, as 16·8 trillion short tons with Asia and European USSR accounting for 65% of these, North America 27%, and Europe and Africa accounting for 5% and 1·5% respectively. Oil shales reserves in the world are given in terms of the grades of shale rock; with grades above 10 gallons of oil per tonne of rock, the world's reserves are 342 trillion barrels of oil as of 1965. Of these, Africa accounts for 25%, Asia for 34%, Europe for 8%, North America for 15% and Latin America for 13%. In the US and Canada about 300 billion barrels of oil are recoverable from the tar sand deposits. It is difficult to quantify and project the world's potential energy yield of cellulosic materials. Solar energy is used not only to grow wood but other plants and crops such as tubers, grasses, etc. However, according to Penner and Icerman[30] about $5·2 \times 10^{24}$ J of mean solar radiation is obtained on the Earth's surface per annum. Of this, about 12×10^{20} J, or 0·023% of total radiation, is used up in photosynthesis. Thus an enormous amount of energy is obtainable from the sun via crop growth. No attempt will be made to estimate the distributive pattern of these radiations. However, on an average crop energy value of $15·4 \times 10^3$ MJ per tonne, the mean solar energy obtainable from photosynthesis is equivalent to about $7·8 \times 10^{10}$ tonnes of crop matter each year. Table 32 indicates that by 1980 about 0·30 cubic metres or 10·6 cubic feet *per capita* of fuelwood will be consumed in the world. At an expected world population of about 4 billion people in 1980 the fuelwood consumption approximates to 1·2 billion cubic metres or 42 billion cubic feet. With an average specific gravity of wood being 40 lb per cubic foot the 1980 consumption rate stands at about 0·95 billion tonnes of wood. The comparison of this figure to the yearly crop growth of mean value of 78 billion tonnes makes it apparently difficult to estimate the percentage of fuelwood in total crop growth rate. An important fraction of grown crops is utilised as food or used in one form or another. However, at a theoretical figure of 0·30 tonne of oil per tonne of cellulosic material liquefied, and assuming that about 20 billion tonnes of grown crops are liquefied, the world can produce an average of 6×10^9 tonnes of oil each year from crop material. This compares favourably with the expected annual oil usage of 6×10^9 tonnes.[30] It may be seen that 100%* of the world's present oil demand can be met by the use of cellulose. Moreover, the

* Based on about 25% of yearly yield of crops.

world's 16·8 trillion short tons of coal can provide about 6 trillion tonnes of oil (based on the ratio of 0·4 tonne oil per tonne coal processed). At a mean rate of consumption of $6·0 \times 10^9$ tonnes per year, this coal-derived oil could last for about 1000 years. From these figures one can see that the present energy demands of the world could actually be maintained from cellulose and coal, not counting the expected oil reserves in the forms of shale rocks and bitumen-impregnated sands. The conclusions to this section are that coal and other fossil fuels can keep the world's domestic energy systems going; there is actually no indication of energy shortages in the future; the only implications and/or problems of turning to less efficient fossil fuels such as coal and cellulose will be those associated with costs. Even so, as will be demonstrated later, the variances between present oil prices and anticipated prices of synthetic fuels will be quite marginal. Thus the overall picture for synthetic fuels is attractive.

12.2.2 The case for methanol use as a motor fuel

The use of methanol as a motor fuel has been a matter of debate in a number of western countries. It is thought that methanol can be used as a motor fuel in possibly two distinct ways. First, it is hoped that methanol can be blended with conventional gasoline, and, secondly, that methanol can be used on its own. Studies done on these modes of utilisation were geared to determine the effects of methanol on fuel economy and emissions, as well as on automobile performance. This aspect of the study concerns mainly institutions dealing with combustion engineering, mechanical engineering etc., and detailed discussion of methanol as a motor fuel is not included. However, because the US is currently investigating gasohol (10% methanol, 90% gasoline), and because many people believe that methanol can aid in motoring in the future, reference is made to Most and Wigg,[57] of Exxon Research and Engineering Company, USA, who give the pros and cons of methanol automotive economy. They report on the performances of vehicles using 15 liquid volume percent methanol–gasoline blends and compared these to vehicles using base gasoline. Two types of blends were tested; one type made from direct addition of methanol to gasoline and the other type made by adjusting the vapour pressure of the blend to the level of the base gasoline. The unadjusted blend obviously had a Reid vapour pressure higher than that of the base gasoline. In adjusting the vapour pressure of the blend all the butane and half of the pentanes of the base gasoline were removed before the methanol was added. In Table 62 some properties of the different types of gasoline–methanol fuel mixtures are presented. Changes obtained in emissions percentages and energy economy

TABLE 62
Composition and Inspection of Fuels for Vehicle Fuel Economy and Emissions Studies

Properties *(components per vol %)*	*Base gasoline*	*Matched RVP[a]*	*Non-matched RVP*
Reformate	58	56	49
Heavy cat. naphtha	0	0	0
Light cat. naphtha	22	22	19
Cat. C-5	13	7	11
Butane	7	0	6
Methanol	0	15	15
RVP (Reid vapour pressure)	11·9	11·7	16·0
ASTM D-86 Dist. IBP	31 °C(88 °F)	36 °C(96 °F)	31 °C(88 °F)
D + L 5%	39 °C(102 °F)	43 °C(110 °F)	33 °C(90 °F)
10%	46 °C(115 °F)	48 °C(118 °F)	39 °C(103 °F)
90%	154 °C(310 °F)	154 °C(309 °F)	152 °C(305 °F)
FBP	189 °C(372 °F)	196 °C(384 °F)	188 °C(371 °F)
RON	98·3	101·7	101·3
MON	86·8	87·5	87·0
API Gravity at 16 °C (60 °F)	54·9	50·2	54·5

[a]Implies that the C-4s and C-5s are partially removed from base gasoline before addition of MeOH.
(Adapted from ref. 57.)

by using methanol – gasoline blends are given in Tables 63 and 64. Essentially the types of engine emissions considered are carbon monoxide, oxides of nitrogen and hydrocarbon fractions. In some cases some aldehydes were emitted.

Tests were also performed on methanol used alone as the fuel. Methanol is believed to have relatively good lean combustion characteristics, and it also has a higher volatility than base gasoline. To extend the studies, methanol – water mixtures were also tested on standard single-cylinder engines. Table 65 gives the octane ratings of methanol – water mixtures and Table 66 shows the properties of methanol and base gasoline (iso-octane). Results of these tests are given more explicitly by Most and Wigg[57] but a summary of the technical conclusions is given here: Table 64 shows that, though the gasoline – methanol blends give net energy gains over base gasoline, the magnitudes of gains do not give enough driving force to necessitate the use of methanol as a blend with conventional gasoline. Table 63 reveals that while the emissions of carbon monoxide and light

TABLE 63
Emission Changes with 15% Methanol Addition to Gasoline

				Emissions *(grams per mile)*			
		CO		*H/C*		*NO$_x$*	
Fuel series	*Car used*	*Base blend*	*15% MeOH*	*Base blend*	*15% MeOH*	*Base blend*	*15% MeOH*
Matched vapour pressure^a	1967	83	41	5·2	3·8	6·4	8·1
	1973	21	8	1·1	1·1	2·6	1·7
	Catalyst	0·3	0·4	0·1	0·1	2·6	2·3
Non-matched vapour pressure^b	1967	83	62	4·4	3·6	5·8	7·7
	1973	18	9	1·1	1·5	2·5	1·9
	Catalyst	1	0·6	0·1	0·2	2·6	2·2

^{a} Matched implies gasoline – methanol blend having some C-4 and C-5 of original gasoline removed.
^{b} Non-matched implies gasoline – methanol blend formed by direct additions.
(Source: ref. 67.)

TABLE 64
Fuel Economy Changes with 15% Methanol Addition to Gasoline

		Fuel economy *(miles per gallon)*		Effect of methanol	
Fuel series	*Car used*	*Base blend*	*15% MeOH*	*Volume basis (%)*	*Energy basis (%)*
Matched Vapour pressure	1967	14·3	14·4	+ 1	+ 8
	1973	11·2	10·5	− 6	+ 1
	Catalyst	11·5	10·9	− 5	+ 2
Non-matched Vapour pressure	1967	14·3	13·5	− 6	+ 1
	1973	12·1	11·4	− 6	+ 1
	Catalyst	11·5	10·8	− 6	+ 1

(Source: ref. 67.)

TABLE 65
Methanol – Water Octane Ratings

Fuel	Research octane number	Motor octane number
Methanol	109·6	97·4
Methanol + 5% vol. of H_2O	110·0	89·5
Methanol + 10% vol. of H_2O	114·0	92·8

(Source: ref. 67.)

TABLE 66
Properties of Methanol and Iso-octane

Property	Methanol (CH_3OH)	Iso-octane (C_8H_{18})
Molecular weight	32	114
Density at 16 °C (60 °F) grams per cm³	0·791	0·692
pounds per gallon	6·62	5·79
Boiling point, °C (°F)	64 (148)	99 (211)
Vapour pressure at 38 °C (100 °F), psia	4·55	1·72
Auto ignition temperature, °C (°F)	470 (878)	447 (837)
Flammable limits, % by volume in air at STP	7·3 – 36	1·1 – 6·0
Specific heat of liquid, Btu per pound at 16 °C (60 °F), 1 atm	0·599	0·489
Heat vaporisation at boiling point and 1 atm, Btu per pound	473	117
Lower heating value, Btu per pound	8 644	19 065
Stoichiometric air – fuel ratio	6·46	15·1
Octane ratings—Motor	87·4	100
Research	109·6	100
Octane blending values—Motor	91	100
Research	120	100

(Source: ref. 67.)

hydrocarbons are decreased with the use of gasoline – methanol blends those of oxides of nitrogen actually increased. Essentially the emission problems remain almost the same. In Table 65 it can be seen that the octane numbers of methanol and methanol – water mixtures are in effect greater than those of base gasoline, and the addition of water to methanol substantially reduces the emission of nitrogen oxides from the car's engine.

With the observations made above pertinent conclusions that could be made are:

(i) On the basis of energy conversion efficiency and emissions from engines, the use of methanol–gasoline blends is not recommendable.

(ii) The combustion characteristics of base gasoline are essentially the same as for methanol–gasoline blends.

(iii) Methanol has a disadvantage of equivalent volume requirement, i.e. to have the same amount of energy as that specified for a certain volume of gasoline about twice the volume of methanol must be used. Moreover, methanol has a higher latent heat of vaporisation and it has a single boiling point characteristic. These properties will make it difficult to achieve acceptable fuel–air mixtures in engines if methanol–gasoline blends are used.

(iv) On the other hand, the use of methanol alone or methanol–water mixtures gives a net energy gain of 34–38% in terms of efficiency of use.

(v) Lower emissions are obtained, compression ratios are higher and octane ratings are higher with a methanol–water fuel than with gasoline.

(vi) Most important of all considerations are the costs per unit of energy used. Assuming methanol and gasoline are made from coal, data presented in Table 48 and Fig. 24 reveal that it will cost about £ 55 per tonne to produce methanol while gasoline can be produced at a cost of £ 59 per tonne. On an energy basis a tonne of gasoline gives about 43 000 MJ while a tonne of methanol gives about 23 100 MJ. Therefore on an end-use basis methanol will cost about 1·5–2·0 times the cost of gasoline. However, if the methanol–water system can be improved upon and enough incentive is obtainable from the results of this system one might anticipate the use of methanol–water fuel in the future. But many parameters come into play in the decision-making of this kind. For example, costs have to be determined more closely, tests have to be done on multi-cylinder engines and possibly the technology of coal-to-methanol has to be improved.

One argument that seems to defeat the future use of methanol is that there are now many routes to oil production, i.e. coal to oil, cellulose to oil, shale rock to oil and bitumen from sand. Essentially these are liquefaction-based processes, and in terms of energy yield the process of liquefaction offers a most attractive path. Methanol production from coal is most economical via the liquid route. Any attempt to provide methanol through gases tends to raise prices. Thus it may not be appropriate to consider methanol as an automotive fuel in the future if gasoline is always readily available.

12.3 SYNTHETIC FUELS ECONOMY AND THE BRITISH SOCIETY

Tables 67 and 68 give the pattern of inland energy consumption in the UK from 1966 to 1978. The annual reports of the British Gas Board,[60] the *Handbook of Electricity Supply Statistics* (1978),[61] the *Electrical Times Electricity Supply Handbook* (1978),[62] the *UK Energy Statistics* (1977)[58] and *Energy Trends*[59] indicate the trend of gas and electrical energy utilisation by both domestic and industrial sectors in the UK. Tables 69–71 indicate energy consumption rates by population or per consumer or per sector in various parts of the country. Percentage increases and/or decreases of the use of various fuels can also be inferred from these tables.

TABLE 67
Total Inland Energy Consumption in the UK, 1966 – 78

Fuels used	1966	1973	1976	Jan.–Nov. 1978
(million tonnes of oil equivalent)				
Coal	104	78·2	71·8	62·6
Petroleum	67·5	96·6	78·9	73·1
Natural gas	0·70	26·0	34·6	32·9
Nuclear electricity	4·7	5·9	7·6	7·1
Hydroelectricity	1·3	1·2	1·1	1·8
Total	178·2	207·9	194·0	177·5
(joules × 10^{15})				
Coal	4 807	3 486	3 141	2 740
Petroleum	3 017	4 365	3 562	3 300
Natural gas	34	1 172	1 558	1 481
Nuclear electricity	190	244	313	292
Hydroelectricity	55	49	47	77

(Source: refs. 58 and 59.)

TABLE 68
Heat Supplied to Final Users (10^{17}J)

	1966	1973	1976	Jan. – Nov. 1978[a]
By fuel				
Coal	18·7	9·4	7·0	5·24
Other solid fuel	8·7	6·2	4·9	3·12
Petroleum products	24·5	34·4	30·4	25·3
Gas	4·3	12·2	15·6	13·9
Electricity	6·0	8·7	8·6	7·20
Total[b]	62·2	70·9	66·5	54·76
By sector				
Iron and steel	8·0	7·7	6·2	4·33
Other industries	18·8	22·3	20·3	16·5
Domestic	16·7	15·0	16·8	14·0
Other consumers	7·6	8·6	8·5	7·11
Total[b]	51·1	53·6	51·8	41·94
Heat per head of population ($J \times 10^{10}$)	14·7	16·6	15·5	13·00

[a] Total fuel consumed by final users in 1977 = $61·00 \times 10^{17}$J; by sectors in 1977 iron and steel consumed $5·15 \times 10^{17}$ J, other industries $18·81 \times 10^{17}$ J, domestic consumers $15·8 \times 10^{17}$ J and other consumers $8·0 \times 10^{17}$ J.

[b] Totals are not the same because sectors such as transport, etc., are not accounted for.

(Source: ref. 59.)

TABLE 69
Electricity Consumption in Selected Areas (MJ(th)[a])

Area	1974 – 75	1975 – 76	1976 – 77
Per head of population			
Manchester	46 492	44 722	44 722
London	41 723	41 446	40 333
Birmingham	41 890	39 176	38 940
Liverpool	39 100	38 114	38 114
Sheffield	82·6	76·7	82·6
Glasgow and Clyde	45 312	43 660	43 424
Edinburgh, Fife and Border	50 976	50 858	50 858
Per head of local consumer[b]			
Manchester	121 131	116 438	116 438
London	92 337	91 028	88 105

TABLE 69 *(Contd.)*

Birmingham	127 386	118 654	117 940
Liverpool	114 356	111 472	111 472
Sheffield	199	188	199
Glasgow and Clyde	119 781	116 161	115 533
Edinburgh, Fife and Border	126 407	124 242	124 242

[a]1 scf of natural gas has an approximate heat value of 1·05 MJ. Therefore calculated values are approximate equivalents of scf of pipeline gas used. Electricity generation is assumed to be at 30% efficiency.
[b]All consumers considered, i.e. both domestic and industrial.

TABLE 70
Electricity Consumption by Regions in Britain (MJ[a])

Region	1975 – 76	1976 – 77	1977 – 78
Per head of population			
Scottish	54 727	56 107	56 932
South eastern	26 790	26 042	27 333
Southern	58 684	58 391	60 738
South western	30 207	30 198	31 407
Eastern	54 696	53 983	55 962
East Midlands	38 010	38 042	39 058
West Midlands	43 803	44 623	46 008
North and south Wales	51 187	54 804	54 731
Northern Ireland	33 923	35 630	36 075
North western	20 590	21 484	21 897
North eastern	65 652	66 330	66 392
Per head of customer			
Scottish	139 088	140 809	141 313
South eastern	94 796	91 123	94 414
Southern	107 177	105 070	107 704
South western	106 913	105 462	108 251
Eastern	104 583	103 220	107 004
East Midlands	113 867	113 962	117 007
West Midlands	124 664	126 997	130 940
North and south Wales	128 384	137 454	137 272
Northern Ireland	103 995	108 476	108 797
North western	111 952	113 109	113 215
North eastern	111 758	116 612	118 850

[a]All values shown are approximate units of scf of pipeline gas equivalent used (1 scf of gas has heat value 1·05 MJ).

TABLE 71
Gas Consumption by Regions in Britain (MJ^a)

Region	1975 – 76	1976 – 77	1977 – 78[b]
Per head of population			
Scottish	12 998	14 322	15 742
South western	13 816	15 514	17 066
South eastern	22 053	21 772	23 949
Southern	19 440	20 773	22 850
Eastern	18 860	19 963	21 960
East Midlands	31 397	33 466	36 813
West Midlands	29 730	31 417	34 560
North and south Wales	21 770	25 434	27 966
North western	28 154	30 152	33 167
North eastern	27 394	29 140	32 054
Per head of customer			
Scottish	78 578	84 868	93 354
South western	81 827	88 216	97 038
South eastern	72 353	70 849	77 933
Southern	83 050	85 832	94 415
Eastern	83 690	85 777	94 354
East Midlands	114 253	118 407	130 248
West Midlands	112 126	115 250	126 774
North and south Wales	112 196	126 638	139 302
North western	100 136	105 146	115 661
North eastern	95 588	99 670	109 637

[a] Gases are usually sold in 100 scf, heat value of 107·3 MJ. Therefore 1 scf of gas is approximately of heat value 1·07 MJ. Hence tabulated values are equivalent scf of gas consumed.
[b] Values extrapolated for 1977 – 78.

Since the sole purpose of synthetic fuels production from solid fossil fuels is to produce the more usable oils and gases, an attempt is made in this section to relate consumption trends with production potentials and, using costs as a basis of comparison, recommend the most economically viable path to solving the impending predicaments of oil and gases shortages. Any recommendations made here are purely on economic grounds though one cannot but remind oneself of the influence of politics and other factors on economics when decisions are to be made on national and international levels. But one thing is certain about meaningful energy policies: it is important that they be based on efficient utilisation of energy. As will be obvious later, the consumption rate of primary heating materials, e.g. coal, gas and oils, in British homes has not changed much over the years. The most likely reason for this is that coal, the most abundant of the fossils in the UK, when it was burnt very inefficiently was rather cheap. However, with the awareness of future scarcity of oil and gas and the rising trend of world inflation, coal is now being used more efficiently. Unfortunately coal today is relatively more expensive.

Table 67 reveals that between 1966 and 1976 the use of coal fell from 58·4% of total fuels used to 37% of the total. Petroleum use rose from less than 40% to about 40·7% of the total. Natural gas rose from less than 1% to 17·8%. Hydro and nuclear electricity rose from 3·4% to 4·5% of the total. Between 1976 and the first 11 months of 1978, coal had another fall of about 2%, petroleum had a rise of 1%, natural gas another rise of 0·7% while electricity had a rise of 0·5%. The pertinent conclusion here is that the use of coal by the final consumer is dropping gradually while the use of gases and oil is rising. It is seen that since 1980 Britain is self-sufficient in oil. Presently petroleum accounts for 41% of total British energy requirements and about 55% of locally produced energy. However, the peak period of petroleum and natural gas is estimated to be around 1990 (Tables 1–4) when oil production will be about ten times its rate as of 1976 and gas production will be about 1·6 times its rate as of 1976. After 1990 it is expected that these will gradually decline in rates of availability to about 1·6 times the level of 1976 for petroleum and 0·13 times the 1976 level for natural gas, in 2025 A.D.

Table 68 reveals that domestic consumers accounted for about 32·4% of the total heat supplied to the final user in 1976 and 33·4% in 1978. Also, the iron and steel industry accounted for 12% of the total in 1976 and 10% in 1978; other industries accounted for 39% of the total in 1976 and 39·3% in 1978. Assuming that the bulk of domestic consumers were predominantly reliant on gas and electricity and a high percentage of industries used petroleum-based heat sources then one can say that coal, petroleum and gas provided the UK with about 83% of its total used energy in 1978. With the decline of oil and gases some other sources of energy have to be tapped to replace them.

Furthermore, Tables 69–71 reveal that, on a basis of scf of gas equivalents, a ratio of 1·7 : 1·0 electricity to gas is estimated per head of population in the UK in 1978. That is to say that as of the 1977–78 period the average Briton used 45 140 scf of gas equivalent by way of electricity consumption and 26 613 scf of gas equivalent by burning natural gas. For the periods of 1976–77 and 1975–76 the ratios were 1·8 : 1·0 and 1·9 : 1·0, respectively. These ratios show a gradual trend towards higher utilisation of gas than electricity. Many other ratios or conclusions can be drawn from

Tables 67–71 but with the above percentages and ratios some relevant conclusions will be attempted here.

It is definite that reserves of oil and gas will gradually be depleted by 2025 A.D. Coal annual production, on the other hand, is expected to rise from 122 million tonnes in 1976 to 170 million tonnes in 2025 (Table 1). Therefore coal is bound to emerge once again in the forefront of British energy economics.

Currently the market price of natural gas sold to consumers can be put as 16·5 pence per therm (anticipated rise of 8% in price by end of 1979) or 0·17 pence per scf of gas. Electricity is sold to consumers, by the North West Electricity Board, at a price of 2·808 pence per kWh(e) or 0·24 pence per scf of gas equivalent of energy. These figures indicate that the average Briton as of 1977–78 was fuelled at a cost of £241·00, based on gas prices, or £340·00 based on electricity prices. On average this means an annual cost of about £291·00 per Briton in terms of total fuelling. Out of this, gas and electricity (Tables 70 and 71) account for an average cost of £147·00 per Briton. This is approximately 50% of the total per head of population costs. Therefore gas and electricity play a very important role in fuelling Britons.

Presently, coal accounts for the major contribution to electricity generation. Overall thermal efficiency of conventional steam stations for generating electricity is about 31·5%. Hydro and nuclear electricity generation is still low in the UK. Thermal efficiency of a typical nuclear station is 26%.[58] On a capital cost basis a nuclear plant will cost about £600 per kW(e) of electricity generated, a coal-fired plant will cost £350 per kW(e), whilst a gas turbine plant will cost £200 per kW(e) (Table 4). Bearing in mind that by 2000 A.D. oil will be scarcer than coal, one can say that the £350 per kW(e) coal plant will be preferable to the gas turbine plant. Also, in terms of abundance the coal-fired plant should be preferred to the £600 per kW(e) nuclear plant. Clearly on an energy economics basis coal can be seen to be playing a large role in electricity generation in future. Also, coal can be shown to be equally economically viable in the gas industries.

In 2000 A.D. the price of coal is expected to be 13 pence per therm (Table 4) or £33·80 per tonne. Using Shearer's coal pyrolysis plus char gasification to pipeline gas process and applying calculations based on this process, as shown in Section 9.6, it can be estimated that synthetic pipeline gas can be produced in 2000 A.D. at a cost of £4·00 per 1000 scf or 40 pence per therm of gas. These calculations are based on 6% inflation each year on both capital and labour cost factors of the process. When compared to the present price of gas it can be argued that a 230% rise in prices of gas (i.e.

costs of production plus expected gains by gas works) in 21 years is not too much to expect. After all the retail price of gas in 1966 was 6·7 pence per therm[58] and in 1979 it is expected to be 17 pence per therm — a rise of about 154% in 13 years. The summary of the argument is that, as of 2000 A.D., pipeline gas can be made from coal and sold at a price of 60 pence per therm. The straight gasification of coal to pipeline gas, as in the Lurgi plant, can yield gas at a production cost of about £ 2·80 per 1000 scf or 30 pence per therm in 2000. Thus a retail gas price of about 52 pence per therm is expected.

The liquefaction process will produce oil from coal, priced at £ 22·80 per tonne, at a mean production cost of about £ 80·00 per tonne of oil based on a plant constructed now or at a cost of £ 180·00 per tonne based on a plant built in 2000 A.D. This will mean that oil produced from coal can sell for about £ 42·00 per barrel or 70 pence per therm in 2000 A.D. The production of oil from cellulose (refuse mainly) in the UK may not be too attractive because of the expected capacity of production. But if used as a supplement to coal-derived oils the net price of oil may be lowered in the future. A 2000 tonnes combustible refuse plant is expected to produce oil of high quality at a cost of about £ 23·00 per tonne of oil in 1978 (Table 60). As of 2000 A.D. this cost of production is expected to be around £ 76 per tonne — based on an annual inflation rate of 6%. Such oils are expected to sell at a price of £ 19·00 per barrel or 31 pence per therm — about 50% of the price of coal-derived oils.

As a net summary the average Briton may have to pay about $2-3\frac{1}{2}$ times what it costs to have the present level of consumption of energy going by 2000 A.D. depending on the source of energy. No attempt has been made to analyse nuclear-based electricity in the future (beyond the scope of this work) and it has been shown that cellulose is bound to have financial advantages over coal as an energy carrier in the future. As a comparison, oil from shale rocks can be obtained at a cost of about £ 50 per tonne of oil in 1978, and oil from bitumen-impregnated sands can be obtained at a cost of £ 56·00 per tonne of oil in 1978. Both figures have been evaluated in Sections 10.1–10.7 and presented in Figs. 25 and 26. These appear to be in the same cost category as oil from coal.

The exploration of the Windscale nuclear waste reprocessing plant and the growth of fuel crops may alter the forecasts made above. Some authorities in other countries, e.g. Hill in the US,[63] believe that the economic salvation in the future lies in the generation of electricity from nuclear energy. For the UK, this type of conclusion is open to debate, as coal is bound to be given some form of recognition in the near future. The recent grant of £ 800 000 by the Government to the National Coal Research Centre at Cheltenham[64] is a gesture, albeit a small one, in this direction. Similar feasibility studies should be encouraged in the cellulose field. The less complex structure of cellulose compared to coal and the relatively lower capital costs of relevant processing plants make cellulose worthy of investigation.

The question now is, how much cellulosic material can Britain provide in the future? Chapter 13 gives an overview of all fossil fuel conversion processes.

CHAPTER 13

An Overview of Solid Fossil Fuel Conversion Processes— A Critical Analysis

13.1 GENERAL COMMENTS

Essentially, Chapters 1–12 attempt to project effects on energy yields and demands of processes converting solid fuels to the more usable liquid and gaseous fuels, and the economic consequences of trying to obtain oils and gases. The higher efficiency of use of the latter fuels cannot be overemphasised. Also in the last chapter we tried to project costs as far forward as 2000 A.D. As a summary of the processes considered the relevant data are given in Table 72.

Looking at the processes, on the basis of costs, cellulose-based processes seem most attractive. Coal liquefaction processes and the tar sands/shale rock processes come next and the pyrolysis of coal leads as regards costs. Looking at product energy yield, most oil-producing processes have products of identical energy value. Methanol processes may have values quite different; in fact half the yield of oil, tonne for tonne. Looking at Table 47 it will be seen that an average coal process requires 3 tonnes of coal to give 1 tonne of oil, 2·9 tonnes of coal to give 1 tonne of methanol, 1 tonne of coal to give 21 000 scf of pipeline gases. 3·3 tonnes of cellulose give 1 tonne of oil and 2 tonnes of cellulose yield 21 000 scf of methane. Also, 12·3 tonnes of shale rock can yield 1 tonne of oil, while 8·5 tonnes of tar sands yield 1 tonne of oil. Thus, in yield consideration, coal and cellulose seem to top the list, followed by bitumen sands and finally shale rock. In terms of world abundance of solid fossil fuels, coal reserves were quoted in 1974 as 16·8 trillion short tons;[30] cellulose can be produced by solar energy at a mean rate of 78 billion tonnes per year; shale oil reserves

TABLE 72

Process	Estimated cost of producing oils/gases at 1978 price levels	Thermal equivalents of products
Direct hydrogenation of coal to oil	£ 72 per tonne of product oil	42 700 MJ per tonne of oil
Pyrolysis of coal to oil and gas	£ 82 per tonne of product oil and 20 pence per therm of gas	42 800 MJ per tonne of oil
Liquefaction of coal to oil by solvent estraction	£ 80 per tonne of product oil	43 200 MJ per tonne of oil
Methanol production from coal	£ 55 per tonne of methanol	23 000 MJ per tonne of methanol
Oil from tar sands	£ 56 per tonne of oil	42 400 MJ per tonne of oil
Oil from shale rocks	£ 50 per tonne of oil	42 700 MJ per tonne of oil
Liquefaction of cellulose to oil	Average £ 24 per tonne of oil	42 900 MJ per tonne of oil
Biological digestion of cellulose to gases	£ 0·80–£ 1·90 per 1000 scf of gas	1000 MJ per 1000 scf of gas
Pyrolysis of refuse to oil, gases and char	£ 2·40 per 1000 scf gas equivalent of products	1000 MJ per 1000 scf gas equivalent of product

for grades above 10 gallons of oil per tonne of rock are 342 trillion barrels or 48 trillion tonnes as of 1965;[30] and bitumen sands are capable of providing 300 billion barrels or 42 billion tonnes of oil. Cellulose is a renewable energy source and would be given the first place in terms of abundance, coal is obviously next, and then oil shale and bitumen sands.

Other factors of production can now be considered. Geographically the growth of cellulose essentially leaves no permanent change on the Earth's surface except during harvesting and planting, whereas coal mining changes the Earth's surface, and its colour causes some environmental changes. The location of oil shale rocks and tar sands makes their exploitation labour and, perhaps, capital intensive. The removal of the

Earth's crust and disposal of spent residue elsewhere brings about permanent changes in geographical structure. In this direction again cellulose leads in consideration. Sociological patterns and traditional ways of life can affect the use of these solid fuels. Most people in developed countries have the benefit of the efficient use of oil and gases. However, various local populations worldwide are used to burning coal, fuelwood or other solid fossil fuels to generate heat depending on which is abundant. But with changing energy trends, the patterns of use should depend on the immediate economically available fuel. One should expect a tropical country like Nigeria to intensify its research into cellulose utilisation and temperate countries such as Britain to develop coal-based technology.

Finally come the political considerations. All statements above neglect the facts that nuclear energy is a strong competitor with domestic fossil fuels, and that other sources of energy such as tidal, wave, etc., are also available. Patience is required on this issue because many nations have differing policies for various reasons: nationalistic, imperialistic, ignorance, and even based on realistic evaluations. But this is the one factor that is most powerful of all in international scenes. The nuclear energy supporter would always condemn burning more coal or oil or gases because of carbon dioxide emission. His fears for the weather changes in future are often overemphasised. The coal-based processor will point at the deadly emissions of plutonium and iodine-131 that may occur because of nuclear processes. He will tend to exaggerate the risk level of such processes. Crop growth has the additional criticism of the low rate of growth and enormous land requirements to be meaningful. However, the fact still remains that some form of policy has to be adopted by each nation. Objectively, one would say that nations with high land capability should grow and use fuelwood, nations with coal should continue to use their coal.

13.2 THE UK SCENARIO

In this section the potentials of coal and other solid fossil fuels are considered with respect to the UK. The case for nuclear power is not considered here since most of the UK's nuclear energy in the future will come from nuclear wastes reprocessing. Moreover, the general discussion of nuclear power is outside the scope of this work. Furthermore, when one calls to mind that between 1974 and 1976 nuclear-based electricity only accounted for about 12% of total electricity utilised (Tables 67 and 73) it will not be improper to concentrate on coal and other resources that accounted for over 70% of the total.

TABLE 73
Methods of Generating Electricity in the UK (GWh)

Methods and countries	1973	1974	1975	1976
Steam (nuclear)				
Total	23 658	29 395	26 518	32 419
England and Wales	21 416	26 928	23 940	28 298
South of Scotland Electricity Board	2 242	2 467	2 578	4 121
Steam (coal and others)				
Total	230 596	216 350	220 301	218 185
England and Wales	202 177	190 428	194 333	193 111
South of Scotland Electricity Board	20 408	18 293	19 484	18 779
North of Scotland Hydroelectricity Board	1 840	1 802	702	281
Northern Ireland	5 467	5 178	5 173	5 324
Gas turbines and oil engines				
Total	1 347	1 139	779	487
England and Wales	1 089	899	460	210
South of Scotland Electricity Board	52	16	9	8
North of Scotland Hydroelectricity Board	187	196	216	242
Northern Ireland	19	28	14	27

(Source: ref. 74.)

Thus essentially the energy sources that will be compared are coal, municipal refuse, specially grown crops and agricultural wastes and oil shale rocks. Factors of comparison will be as follows:

  (i) Energy requirements of processes that are conventionally used to produce oil and gases from the solid fuels. In this aspect of consideration the energy demands – recoverables, as calculated in Part I, will be used as scaling factors.
 (ii) Costs estimates calculated in this section will be considered and relative attractiveness will be apportioned to the fossil fuels concerned.
(iii) Geographical factors such as location, distribution and environmental disturbances.
 (iv) Modes of collection and associated ease or difficulties.
  (v) The practical aspect of potential contributions of each fuel to the UK's annual energy consumption.

(vi) Abundance and hence continuity of supplies.
(vii) A mathematical representation of the factors given above will be made.

From Table 46 and Fig. 23 it can be seen that the liquefaction of coal to synthetic oil has a net energy recovery – demand ratio of 0·60. This means that if dried and prepared coal is extracted with solvent and hydrogenated to yield 1 tonne of usable oil product, about 1·68 tonnes of the product oil equivalent is required as input energy to the conversion process leaving a net yield of 0·32 tonne (effective) of oil as gain in the process. In another form this implies that to get 1 tonne of oil from coal by conversion techniques about 1·68 tonnes oil equivalent of energy is required for the liquefaction process. Thus the recovery – demand ratio is 1·00 : 1·68 or 0·6 : 1·0. Pyrolysis of coal to oil and gasification of resulting char to pipeline gas gives a net energy recovery – demand ratio of 1·0 : 1·8 or 0·55 : 1·0. The Fischer – Tropsch's methanol synthesis process yields a recovery – demand ratio of 0·56 : 1·0. The extraction of oil from bitumen-impregnated sands yields a recovery – demand ratio of 1·27 : 1·0. Extraction of oil from oil shale rocks has a recovery – demand ratio of 3·30 : 1·0. Liquefaction of municipal refuse to usable oils has a recovery – demand ratio of 1·44 : 1·0. Liquefaction of specially grown plants, and associated agricultural wastes, to usable oils has a recovery – demand ratio of 2·45 : 1·0. Biological digestion of crop matter (grown plants) to pipeline gases has a recovery – demand ratio of 2·23 : 1·0. Finally the pyrolysis of refuse to oil, gases and residual char has a recovery – demand ratio of 0·43 : 1·0. Thus with energy demands being reduced to a common factor of 1·0 the relative attractiveness of the processes discussed puts the parent fossil fuels in the positions shown in Table 74.

TABLE 74

Fossil fuel	*Potential energy recovery capability*[a]	*Relative position*
Coal	0·60	4
Refuse	1·44	3
Grown crops	2·30	2
Oil shale	3·30	1

[a] Simple ratios, no units.

TABLE 75
Costs of Production of Synthetic Fuel Energy from Solid Fossil Fuels

Solid fossil fuel conversion process	Product from process and costs of production	Heat value of product	Cost of producing 1 GJ of energy
Coal liquefaction to syncrude	£ 80·00 per tonne of synthetic crude	43 000 MJ per tonne	£ 1·85
Coal pyrolysis to oil and gas	£ 82·00 per tonne of syncrude and 20 pence per therm of gas	Synthetic oil is 43 000 MJ per tonne; 1 therm is 105 MJ	£ 1·90[a]
Extraction of oil from shale rocks	£ 50·00 average per tonne of oil	42 700 MJ per tonne	£ 1·17
Liquefaction of refuse to oil	£ 26·00 per tonne of oil	43 000 MJ per tonne	£ 0·60
Liquefaction of grown crops to oil	£ 22·00 per tonne of oil	43 000 MJ per tonne	£ 0·50
Biological digestion of crops to gas	£ 1·20 average per 1000 scf of gas	1050 MJ per 1000 scf	£ 1·15
Pyrolysis of refuse to synthetic fuels	£ 2·35 per million Btu of product gas, oil and char	1000 MJ of product	£ 2·35

[a] Ratio of oil to gas yields in process in terms of energy = 0·7:1·0.

TABLE 76

Fossil fuel	Cost of producing 1 GJ synthetics	Relative position
Coal	£ 1·67–£ 1·90	2
Refuse	£ 0·60 (liquefaction)	1
	£ 2·35 (pyrolysis)	3
Oil shale	£ 1·20	2
Grown crops	£ 0·50	1

TABLE 77

Fuel	Relative importance with respect to geographical considerations
Coal	3
Refuse	1
Grown crops	2
Oil shale	4

Costs of production of synthetic fuels will be based per GJ or 10^6 Btu of energy yield. From cost data already calculated the figures shown in Table 76 could be extracted from Table 75.

Geographical factors are a matter of subjective judgement. There is no real way of quantifying environmental disturbances, location attractiveness and the rest. However, the ability to derive oil from refuse should give it some form of exaggerated credit. Local populations will continue to dispose of wastes and if these wastes can be gainfully utilised then one should give refuse first position (Table 77). Mining for both coal and oil shale rocks permanently causes geographical patterns. About 90% of the UK's coal is won by deep-mining[65] and oil shale has to be taken off to retorting sites. Oil shale has the additional problem of spent rocks disposal.

With regard to mode of collection, ease or difficulty of collection and associated factors, coal and refuse tend to vie for first place. Coal mining technology is so advanced in the UK that its collection system is similarly advanced. Rail haulage is the main mode of collection from collieries to users, and the cost of transportation of coal by rail is about 20 pence per tonne.[66] Refuse collection averages £1·60–£1·80 per tonne (Appendix C6), a cost based on the present mode of collection by special vehicles. However, the Greater London area is planning rail haulage for refuse.[67] With this fully introduced, refuse collection could be quite similar in costs to that of coal. Presently coal collection costs can be given first place. Most agricultural crops are grown in rural farmlands and any attempt to collect these has to follow the conventionally used pattern, i.e. the use of tractors and trucks, making grown crops collection just as expensive as that of refuse. However, refuse collection will in future have an edge over grown crops because the former could be more organised than the latter. As is evident in Tables 54–57 the processing of oil shale rocks is labour intensive, and some of this labour is in the transportation. Hence the relative

TABLE 78

Fuel	Relative position with respect to mode of collection
Coal	1
Refuse	2
Grown crops	2
Oil shale	3

attractiveness of the processes with regard to collection is shown in Table 78.

What is the potential energy contribution of these fossil fuels to the total UK energy demand? It has been noted that the UK has a reserve potential of 1006 million barrels or 144 million tonnes of oil from its oil shale rock deposits.[51] The national energy demand of the UK in 1978 was about 194 million tonnes of oil equivalent. Thus, assuming that all shale oil is recoverable (not true) and that the pattern of energy consumption does not change over the next 10 years, and furthermore that all the energy demands of the nation are to be obtained from shale oil (fictitious), it can be seen that the oil shale deposits cannot serve the UK's need for even a year.

Refuse capacity of Britain is put by Schomburgh[33] as 16·5 million tonnes annually (rather low), based on both domestic and commercial outputs. On the domestic scene alone it is fairly difficult to quantify exact outputs. However, from the summary of refuse disposal statistics 1978–79, prepared by the Greater Manchester County Council, the refuse disposal trend shown in Table 79 can be obtained. Statistically the figures presented

TABLE 79

County areas	Refuse disposal per head of population (tonnes per year)
Greater Manchester	0·84
West Midlands	0·35
Tyne and Wear	0·44
West Yorkshire	0·73
South Yorkshire	0·68
Merseyside	0·46

in Table 79 are not quite representative of the UK. But using an average of 0·5 tonne per head of population of 56 million, one can estimate a domestic refuse disposal of about 28 million tonnes. Suppose the combustible materials are put at about 50–60% of total refuse and that the yield per tonne of combustible material is 0·3 tonne of oil. The net recovery from domestic refuse annually will be approximately 5 million tonnes of oil. This is equivalent to about 2·6% of total national demand each year. With grown plants the problem is one of land restriction. The solar constant* of the UK is about 110 W per square metre. This is a third of the mean world value of 340 W per square metre. That is to say that the UK sunshine is just one-third of the sunniest region in the world. Based on this fact Cooper calculated the potential and actual photosynthetic production in the UK;[68] total solar energy input was $7·0 \times 10^{14}$ MJ in 1974. At a 1% conversion of total radiation, potential photosynthetic production on both cultivable land and rough grazings was put at $7·0 \times 10^{12}$ MJ. But the actual photosynthetic production was calculated as $1·1 \times 10^{12}$ MJ in 1975. This was barely 16% of the potential yield. The UK had a total land area of 24 102 000 hectares in 1976, and in the same year the total agricultural area was 18 990 000 hectares. Of this agricultural land, food crops accounted for about 26%, grass and grazings accounted for 72% while fuelwood and others accounted for the remaining 2%.[65] The question now is how much of these can be converted to an energy crop? Certainly not all parts of food crops are directly consumed by man, e.g. sugar beet has about 50% as the leafy material not consumable by man. Suppose we now assume that agricultural authorities are ready to give 25% of grassland to energy yield each year and about 40% of food crops are laid waste as non-edibles (Cooper's estimates are in close order), then a net equivalent of 5 400 000 hectares of land will be available to cultivate grass purposely for conversion to oil. From calculations done in Section 11.5, about 13 200 hectares will be needed to yield about 1000 tonnes of dry grass per day; 5 400 000 hectares will give 0·41 million tonnes per day or 106 million tonnes per year (assuming 260 working days). With cellulose-oil conversion efficiency of 30% an expected annual oil yield of 32 million tonnes can be calculated. Applying a 75% accuracy factor on this the UK can produce about 24 million tonnes of oil from grown grass annually. This is approximately 12·5% of current national energy demand.

   In 1970 the National Coal Board put the UK's coal reserves as 4200 million tonnes of economically workable and 3000 million tonnes po-

---

* Solar constant is the intensity of solar radiation on an area of the Earth.

TABLE 80

Fossil fuel	Relative positions with respect to abundance and percent contributions
Coal	1
Refuse	3
Grown crops	2
Oil shale	4

tentially available,[69] and between 1970 and 1978 an average of 147 million tonnes of coal have been used each year.[65] On the basis of reserves quoted above, the national present expected reserves are about 6000 million tonnes. Suppose recovery potential is put at 90%, i.e. 90% of reserves are actually recovered, then about 5400 million tonnes can still be used to generate oil or gases. If the same basis of oil output is applied, as in the case of other sources, this will mean a total reserve of 1860 million tonnes of synthetic oil recovery. Thus, if coal alone were to support the British energy system, another 10 years of supply is expected. Note that these evaluations are based on the fact that the resulting char residues are not utilised. This is far from being true and the life expectancy of UK coal reserves is much more than that calculated. Presently, coal supplies about 35% of the national annual energy demand (Table 67).

In the net analysis, if we restrict our timing to about 10 years from now, the relative position of the fossil fuels can be given as shown in Table 80. Positions may change in time because coal will be ever depleting while refuse and grown crops are renewable sources.

13.3 ATTRACTIVE RATIOS OF SOLID FOSSIL FUELS AS SOURCES OF LIQUID AND GASEOUS SYNTHETIC FUELS

As a form of mathematical summary which will enable one to recommend the various solid fuels in order of their merits with regard to their future contributions in synthetic fuel production in the UK, this section re-evaluates the factors of production already dealt with in Section 13.2. As a recap, these factors are costs of producing synthetics from solid fuels, the expected energy recovery – demand ratios, geographical impacts of the exploration of these solid fuels, the modes of collection of these fuels and

TABLE 81

Attractiveness factors[a] for Solid Fuels as Sources of Liquid and Gaseous Fuels in the UK

Factors considered	Ranked weighting	Coal		Refuse		Grown grass		Oil shale	
		Relative points[b]	Total points[b]	Relative points[b]	Total points[b]	Relative points[b]	Total points[b]	Relative points[b]	Total points[b]
Energy recovery–demand ratio of conversion process	20	1	20	2	40	3	60	4	80
Cost of producing 1 GJ of energy equivalent of synthetic fuel	18	3	54	4	72	4	72	3	54
Abundance and relative contribution of yield to national energy demands	50	4	200	2	100	3	150	1	50
Modes of collection of solid fuels	7	4	28	3	21	3	21	2	14
Geographical and sociological impact of process	5	2	10	4	20	3	15	1	5
Grand total	100		312		253		318		203

[a] Purely mathematically based.
[b] Relative points given in reverse order of positions, 1–4. Total points = relative points multiplied by weight factor.

finally the relative abundance and contributions of products to the UK's energy demands. Obviously in considering these factors the last seems to have more significance than any of the others. If a solid fuel is not available it will be unnecessary to consider costs or recovery efficiencies or impact on local population. Next in significance is yield: a high yield tends to compensate for higher costs. However, almost equally important is the cost of production. Modes of collection can be placed next and finally geographical–sociological restrictions. An attempt is made here to 'weight' these factors objectively. Abundance of solid fuel and the percentage contributions to total national demand will be given 50 out of a total 100, energy yield from solid fuel 20, costs 18, collection 7 and geographical–sociological impact 5. Essentially one is saying that in terms of importance to energy economics their ratios are roughly $10 : 4 : 3.6 : 1.4 : 1.0$.

With these weight factors Table 81 is constructed, and from the results obtained it can be said that cellulose and coal have a promising future in the UK. Provided about 25% of present grasslands and crops wastes are set aside as sources of energy, cellulose has a primary part to play in the production of synthetic fuels in the UK in the future. Furthermore, refuse can provide additional inputs.

Thus essentially the UK can afford to fuel each person in 2000 A.D. at a cost of about 40–60 pence per therm or £710 per annum and still maintain the present level of energy consumption which is equivalent to 3.40 tonnes of petroleum per head of population per year. All that is needed are realistic energy policies which, in our opinion, should consider coal and cellulose-based conversion processes closely. Appendix C9 discusses the projected and present relative costs of nature-made petroleum and natural gas, the costs of producing oils and gases from coal and costs of producing oils and gases from cellulosic materials.

APPENDICES

Evaluation of GER for H-Coal Process

USING EASTERN COAL OF ILLINOIS

The sequence of reactions and general characteristics of coal fed into the reaction system are to be found in a reference by Goen *et al.*[70]

Inputs

Coal

Goen *et al.*[70] estimate that for a plant producing 14 452 tonnes of syncrude per standard day the total requirements of the Illinois-type coal would be 34 767 tonnes per standard day. This would be equivalent to 30 300 tonnes of dry coal per standard day. Hemming[11] carried out

TABLE 82

	Quantity of moisture-free coal (tonnes per standard day)	% of total
Direct coal converted to syncrude (theoretical minimum)	17 997	57·5
Hydrogen production (theoretical minimum)	10 804	34·5
Char byproduct	530	1·7
Plant fuel and sundry losses	1 969	6·3
	31 300	100

detailed mass and energy balances for the elements carbon, hydrogen, oxygen and nitrogen present in the coal and arrived at the coal equivalent inputs shown in Table 82.

Electricity

Goen *et al.*[70] assume in the conceptual design that annual electricity required for this process is of the order of $10\cdot304 \times 10^8$ kWh(e). If it is assumed that this quantity of electricity is generated on-site by a coal-fired plant operating at 30% efficiency, with no transmission losses, and that the plant is on-stream for 90% of the time, then energy requirements for electricity

$$= 10\cdot304 \times 10^8 \times 3\cdot6 \times \frac{100}{30} \times \frac{1}{365} \times \frac{100}{90}$$
$$= 0\cdot376 \times 10^8 \text{ MJ(th) per standard day}$$

Note: (i) 1 kWh(e) is taken as $3\cdot6 \times 10^6$ J(e)
        (ii) In coal equivalent, electricity used will be

$$\frac{0\cdot376 \times 10^8}{\text{CV of coal}}$$
$$= \frac{0\cdot376 \times 10^8}{25\,530}$$
$$= 1473 \text{ tonnes of coal}$$

Coal slurry preparation

Additional coal is required for the preparation of coal-derived solvent. This solvent is used to prepare the 40-mesh ground coal into a slurry. This additional quantity was estimated by Hemming[11] to be about 4686 tonnes equivalent of coal but was not included in the energy required because the solvent is recycled.[70] Make-up solvent is required to account for losses.

Plant and capital equipment

Goen *et al.*[70] estimate the cost of an H-coal plant producing 100 000 barrels per standard day, i.e. 14 452 tonnes of syncrude per standard day, as

$(US 1974) 533 million

But

$(US 1974) 1\cdot0 = \$(US 1968) 0\cdot74,$

as shown by the US Department of Commerce, Bureau of the Census[71] and

$$\$ (US\ 1968)\ 2\cdot4 = £ (1968)\ 1\cdot0$$
$$\$ (US\ 1978)\ 1\cdot85 = £ (1978)\ 1\cdot0$$

Therefore estimated cost of plant as of 1978

$$= £ (1978)\ 533 \times 10^6 \times 0\cdot74 \times \frac{1\cdot85}{2\cdot4}$$

$$= £ (1978)\ 304 \text{ million}$$
$$= £ (1968)\ 164\cdot3 \text{ million}$$

Also, Casper *et al.*[72] give the energy intensity of the products of a number of heavy engineering industries, e.g. steel, as about 220 MJ(th) per £ (1968). Thus if it is assumed that the plant is on-stream for 90% of its 20-year lifetime, and 50% of its initial costs are allowed for maintenance, then energy requirement of plant and equipment

$$= 1\cdot5 \times 164\cdot3 \times 10^6 \times 220 \times \frac{1}{20 \times 265 \times 0\cdot9}$$
$$= 0\cdot083 \times 10^8 \text{ MJ(th) per standard day}$$

In terms of coal equivalent, that is

$$\frac{0\cdot083 \times 10^8}{25\ 530} = 325 \text{ tonnes per standard day}$$

Catalysts and chemicals
Goen *et al.*[70] estimate the annual cost of catalysts and chemicals as $ (US 1974) 10·73 million. From US wholesale price indices[71] for chemicals and allied products

$$\$ (US\ 1974)\ 1\cdot0 = \$ (US\ 1968)\ 0\cdot68$$

and using the same procedure as before, noting that the energy intensity for general chemicals, as given by Casper *et al.*,[72] is 702 MJ(th) per £ (1968), we can calculate the energy requirement of chemicals and catalysts as

$$10\cdot37 \times 10^6 \times \frac{0\cdot68}{2\cdot4} \times 702 \times \frac{1}{365 \times 0\cdot9}$$

$$= 0\cdot063 \times 10^8 \text{ MJ(th) per standard day}$$

or in coal equivalents

$$= \frac{0 \cdot 063 \times 10^8}{25\,530}$$

$$= 247 \cdot 0 \text{ tonnes per standard day}$$

Water requirements

Goen *et al.*[70] give the following water requirements but they are not included in NER calculations (the electrical power used by pumps, etc., has been included in the electricity section):

cooling water $2 \cdot 84 \times 10^6$ tonnes per standard day
boiler feed water $2 \cdot 30 \times 10^4$ tonnes per standard day
steam $3 \cdot 06 \times 10^4$ tonnes per standard day

Therefore total coal input (as received) = 34 767 for actual operation + 1473 for electricity

$$= 36\,240 \text{ tonnes per standard day}$$

Hence GER of process

$$= 36\,240 + 325 + 247$$
$$= 41\,598 \text{ tonnes coal equivalent per standard day}$$
$$= 41\,598 \times 25\,530 \text{ MJ(th) per standard day}$$
$$= 10 \cdot 583 \times 10^8 \text{ MJ(th) per standard day}$$

Outputs

Goen *et al.*[70] give the following products and production rates for the conceptual design:

syncrude	14 452 tonnes per standard day
coal char	368 tonnes per standard day
ammonia	378 tonnes per standard day
sulphur	925 tonnes per standard day

Following the convention of IFIAS, the energy requirements of this process are partitioned between fuel products according to calorific values and between non-fuels according to replacement energy requirements. Harris[73] gives the following data as the calorific values and replacement energies of the products:

syncrude	4·27 GJ(th) per tonne
coal char	33 GJ(th) per tonne
(100% carbon)	

ammonia   55·71 GJ(th) per tonne
sulphur   400 MJ(th) per tonne

Therefore % of total plant energy requirements that can be apportioned to syncrude manufacture

$$= \frac{14\ 452 \times 42\ 700}{(14\ 452 \times 42\ 700) + (368 \times 33\ 000) + (378 \times 55\ 710) + (925 \times 400)}$$

$$= 0·949$$

The thermal efficiency of the process is given by the ratio

$$\frac{\text{total calorific value of products}}{\text{total calorific value of inputs}}$$

Thermal efficiency of process

(a) with respect to syncrude only $= \dfrac{14\ 452 \times 42\ 700}{9·398 \times 10^8}$

$$= 61·0\%$$

(b) with respect to all products $= 64·3\%$

USING WESTERN COAL OF WYOMING RIVER

Inputs
Following a similar treatment as above, the summary of total energy inputs is given in Table 83.

TABLE 83

	10^8 MJ(th) per standard day
Coal for syncrude production	4·611
Coal going to char byproduct	0·099
Coal for hydrogen production	3·454
Coal for other plant fuel and sundry losses	0·915
Coal for electricity generation	0·410
Plant and capital equipment	0·082
Catalysts and chemicals	0·072
	9·643

Outputs

According to design, as in ref. 70, the following were the products:

> 13 538 tonnes of syncrude per standard day (CV = 42 700 MJ(th) per tonne)
>
> 248 tonnes of char per standard day (CV = 33 000 MJ(th) per tonne)
>
> 188 tonnes of sulphur per standard day (RV (replacement value) = 400 MJ(th) per tonne)
>
> 289 tonnes of ammonia per standard day (RV = 55 710 MJ(th) per tonne)

Therefore % of total plant energy requirements apportioned to syncrude

$$= \frac{13\ 538 \times 42\ 700}{(13\ 538 \times 42\ 700) + (248 \times 33\ 000) + (289 \times 55\ 710) + (188 \times 400)}$$

$$= 0 \cdot 960$$

Thermal efficiency of process

(a) with respect to syncrude only $= \dfrac{13\ 538 \times 42\ 700}{9 \cdot 644 \times 10^8}$

$$= 60 \cdot 0\%$$

(b) with respect to all products $= 62 \cdot 5\%$

Evaluation of GER for the CSF Process

USING EASTERN COAL OF ILLINOIS

The Consol Synthetic Fuels (CSF) process uses the hydrogen-donor principle to produce a de-ashed extract which is subsequently upgraded by catalytic hydrogenation (cobalt – molybdenum catalyst at about 450 °C and 3000–4200 psi) to yield a medium quality synthetic crude suitable as a feedstock for a conventional refinery. Hemming[11] presents some calculations of the GER of this process based on the summary report of 'Project Gasoline' as carried out by the Consolidation Coal Company (CCC),[16] which are shown below.

Inputs

Coal
CCC[16] estimates that 21 800 tonnes of dry coal per day would be required to fuel the plant. This quantity is distributively utilised as follows:

1486·4 tonnes gasified and used as plant fuel;
about 8350 tonnes converted into char and used also for fuelling plant;
the remainder is used for syncrude production.

Electricity
The estimated daily electricity demand is about $1·552 \times 10^6$ kWh(e). Suppose electricity plant is on-site and on-stream 90% of the time at 30%

181

efficiency. In tonnes of coal equivalent, this energy demand is equal to

$$1.552 \times 10^6 \times 3.6 \times \frac{100}{30} \times \frac{100}{90} \times \frac{1}{27\,170}$$

$$= 685.5 \text{ tonnes of dry coal per standard day}$$

Catalysts and chemicals
Estimated annual cost of chemicals and catalysts for plant is $(1969) 7.52 million. From US wholesale price indices[71] for chemicals and allied products

$$\$(1969)\,1.0 = \$(1968)\,1.0$$

and

$$\$(1968)\,2.4 = £(1968)\,1.0$$

Also, Casper[72] gives the energy intensity of 'general chemicals' as 702 MJ(th) per £ (1968). So, assuming that the plant is on-stream for 90% of the time, the energy requirement of catalyst and chemicals

$$= 7.52 \times 10^6 \times \frac{1.0}{2.4} \times 702 \times \frac{1}{365 \times 0.9}$$

$$= 6.70 \times 10^6 \text{ MJ(th) per standard day}$$

Capital equipment
The cost of the plant and equipment is estimated at $(1969) 239.6 million. From ref. 71, again, the wholesale price indices for the machinery and equipment sector give

$$\$(1969)\,1.0 = \$(1968)\,0.97$$

and

$$\$(1968)\,2.4 = £(1968)\,1.0$$

Casper *et al.*[72] give the average energy intensity of the outputs of a number of heavy engineering sectors (i.e. energy contents of equipment) as 220 MJ(th) per £ (1968).

Assuming the plant is on-stream 90% of its 20-year lifetime and that 50% of initial costs are allowed for maintenance, the energy requirement of plant and equipment

$$= 1.5 \times 239.6 \times 10^6 \times \frac{0.97}{2.4} \times 220 \times \frac{1}{20 \times 365 \times 0.9}$$

$$= 4.86 \times 10^6 \text{ MJ(th) per standard day}$$

Therefore total GER of process

$$= (21\ 800 \times 27\ 170) + (685 \cdot 5 \times 27\ 170) + (6 \cdot 70 \times 10^6) + (4 \cdot 86 + 10^6)$$
$$= 6 \cdot 224 \times 10^8 \text{ MJ(th) per standard day}$$

Outputs
Expected daily products are

7667·3 tonnes of motor spirit (CV = 46 830 MJ(th) per day)
32·3 tonnes of phenol and cresols (RV = 50 000 MJ(th) per day)
176·6 tonnes of ammonia (RV = 55 710 MJ(th) per day)
555·8 tonnes of sulphur (RV = 400 MJ(th) per day)

% of total plant energy requirements apportioned to motor spirit

$$= \frac{7667 \cdot 3 \times 46\ 830}{(7667 \cdot 3 \times 46\ 830) + (32 \cdot 3 \times 50\ 000) + (176 \cdot 6 \times 55\ 710) + (555 \cdot 8 \times 400)}$$

$$= 0 \cdot 968$$

Thermal efficiency of process
(a) with respect to motor spirit = 57·5%
(b) with respect to all products = 59·4%

USING WESTERN COAL OF MONTANA

Calorific value of dry coal = 27 360 MJ(th) per tonne

Inputs

Coal
About 21 469 tonnes of dry coal required.

Electricity
About 744 tonnes of dry coal per standard day required to generate this.

Chemicals and catalysts
Annual costs of these estimated at about $ (1969) 6·37 million. This, using methods as above, turns out to be equivalent to energy requirement of

$$5 \cdot 24 \times 10^6 \text{ MJ(th) per standard day}$$

Total GER of process = $619 \cdot 5 \times 10^6$ MJ(th) per standard day

Outputs
Expected daily products are

> 7607 tonnes of motor spirit (CV = 46 830 MJ(th) per day)
> 32·8 tonnes of phenol and cresols (RV = 50 000 MJ(th) per day)
> 139·3 tonnes of ammonia (RV = 55 710 MJ(th) per day)
> 122·2 tonnes of sulphur (RV = 400 MJ(th) per day)

% of total plant energy requirements apportioned to motor spirit

$$= \frac{7607 \times 46\,830}{(7607 \times 46\,830) + (32·8 \times 50\,000) + (139·3 \times 55\,710) + (122·2 \times 400)}$$

$$= 0·974$$

Thermal efficiency of process
(a) with respect to motor spirit only = 57·5%
(b) with respect to all products = 59·4%

Evaluation of GER for COED Processes

PYROLYSIS WITH CHAR GASIFICATION

The COED (Char–Oil Energy Development) process was developed by the FMC corporation under contract to the office of coal research of the US Department of Interior. The process is a multi-stage fluidised bed pyrolysis producing oil, gas and char. Dried, crushed coal is treated in four fluidised bed stages at successively higher temperatures until a major fraction of the volatile matter of the coal is obtained by burning a portion of the char with oxygen in the last stage.

Shearer,[17] in his conceptual design and economic study for a plant, combines the COED pyrolysis process with a low-pressure version of the Kellogg molten salt process of gasifying char. Hemming[11] carried out calculations based on Shearer's work and the summary is given below.

Inputs

Coal
For a plant producing $7·08 \times 10^6 \text{ m}^3$ of pipeline gas and 27 275 barrels of synthetic crude oil per standard day, the daily coal throughput is estimated to be 28 455 tonnes (calorific value = 26 040 MJ(th) per tonne). Distributive use of this quantity of coal is as follows:

    4013 tonnes for generating electricity, estimated to be 362·8 MW
     895 tonnes burnt in furnace to supply heat for first stage of pyrolysis and coal drying

1448 tonnes for use to provide on-site facilities such as process steam, feed water for boilers, etc.

22 099 tonnes for syncrude and pipeline gas

Hydrogen formation is done by the reforming of product gas. Also the portion of coal burnt to provide heat requires about 1820 tonnes of oxygen for combustion. Therefore

$$\text{input energy from coal} = 28\ 455 \times 26\ 040$$
$$= 7.41 \times 10^8 \text{ MJ(th) per standard day}$$

Capital equipment
Estimated cost of plant is $(1973) 439 million. Using the approach as in Appendix A2, the energy requirement due to equipment is estimated to be 7.81×10^6 MJ(th) per standard day.

Catalysts and chemicals
Shearer[17] estimated that the requirement for sodium carbonate (for the molten salt gasifier) would be 564 tonnes per standard day, and the annual costs of other chemicals and catalysts are estimated to be $(1973) 2.45 million. Also using the approach as in Appendix A2, the estimated energy requirements due to catalysts and chemicals $= 11.66 \times 10^6$ MJ(th) per standard day. Therefore

$$\text{total GER of process} = 7.605 \times 10^8 \text{ MJ(th) per standard day}$$

Outputs
Expected daily products in the conceptual design are

27 275 barrels of syncrude (CV = 6120 MJ(th) per barrel)
7.08×10^6 m³ pipeline gas (CV = 34.31 MJ(th) per cubic metre)
1900 barrels of light hydrocarbons (CV = 4120 MJ(th) per barrel)
941 tonnes of sulphur (RV = 400 MJ(th) per tonne)
36 tonnes of phenol (RV = 50 850 MJ(th) per tonne)

% of total plant energy requirement apportioned to syncrude production

$$= \frac{27\ 275 \times 6120}{\begin{array}{c}(27\ 275 \times 6120) + (7.08 \times 10^6 \times 34.31) + (1900 \times 4120) \\ + (941 \times 400) + (36 \times 50\ 850)\end{array}}$$

$$= 39.75\%$$

Thermal efficiency of process
(a) with respect to syncrude only $= 22 \cdot 0\%$
(b) with respect to all products $= 55 \cdot 2\%$

PYROLYSIS WITHOUT CHAR GASIFICATION

Eddinger,[18] in an economic examination of the commercial plant in Utah producing crude oil, pipeline gas and char for sale, provided a conceptual design for a plant based only on pyrolysis.

Inputs

Coal
According to Eddinger,[18] 9091 tonnes of dry coal per standard day will be required. The coal as received had a moisture content of 6% and this would imply that the throughput of coal, as received, would be 9636·4 tonnes per day. Also, by design, the plant would only work for 330 days per year. The energy requirement, from coal input, assuming CV of dry coal $= 31\ 400\ \text{MJ(th)}$ per tonne would be

$$9636 \cdot 4 \times 31\ 400 \times \frac{100}{106} = 285 \cdot 5 \times 10^6\ \text{MJ(th) per standard day}$$

Oxygen production
Estimated oxygen requirement (for combustion purposes), by design, is put at 794·3 tonnes per standard day. The extra energy requirement* for production of this is estimated as

$$794 \cdot 3 \times 6330 = 5 \cdot 03 \times 10^6\ \text{GJ(th) per standard day}$$

Electricity
By design, the estimated electricity requirement is 574 400 kWh(e). Suppose this is generated on-site from coal at 30% efficiency, then the extra coal to be used will be about

$$574\ 400 \times 3 \cdot 6 \times \frac{100}{30} \times \frac{1}{31\ 400} = 219 \cdot 5\ \text{tonnes dry coal per standard day}$$

This is equivalent to an energy requirement of

$$6 \cdot 89 \times 10^6\ \text{MJ(th) per standard day}$$

* Energy requirements per tonne of oxygen produced $= 6330\ \text{GJ(th)}$ per tonne.

Capital equipment
Estimated plant cost is $(1970) 32·70 million. Using the approach as in Appendix A2, an energy requirement of $0·63 \times 10^6$ MJ(th) per standard day is calculated.

Note: There is no need for catalysts.
Therefore

$$\text{total GER} = (0·63 + 6·89 + 5·03 + 285·5) \times 10^6$$
$$= 298·05 \times 10^6 \text{ MJ(th) per standard day}$$

Outputs
Expected products from process are

$2028·6$ tonnes of syncrude (CV = 43 000 MJ(th) per tonne)
$4781·8$ tonnes of char (CV = 19 520 MJ(th) per tonne)
$1·10 \times 10^6$ m³ gas (CV = 21·46 MJ(th) per cubic metre)

% of total plant energy requirement apportioned to syncrude production

$$= \frac{2028·6 \times 43\ 000}{(2028·6 \times 43\ 000) + (4781·8 \times 19\ 520) + (1·10 \times 10^6 \times 21·46)}$$

$$= 0·427 \text{ or } 42·7$$

Thermal efficiency of process
(a) with respect to syncrude only = 29·3%
(b) with respect to all products = 82·6%

Evaluation of GER for Fischer – Tropsch Processes

MOTOR SPIRIT PRODUCTION

Chan et al.[19] prepared a conceptual design of a plant producing primarily motor spirit, this being based on actual operating data from the Sasol plant in South Africa. Their report was used for the energy analysis shown below, as presented by Hemming.[11]

Inputs

Coal
For a plant producing 44 500 barrels of motor spirit and other hydrocarbons per standard day, Chan estimated the coal input as 31 135 tonnes per standard day (CV = 20 600 MJ(th) per tonne). 40% of this feed is used for stream and power generation. Therefore

$$\text{energy input from coal} = 31\ 135 \times 20\ 600$$
$$= 641 \cdot 3 \times 10^6 \text{ MJ(th) per standard day}$$

Plant and capital equipment
Estimated cost of the Sasol-type plant was $(US 1975) 533 million. From US wholesale price indices[71] for machinery and equipment sector

$$\$(\text{US 1975})\ 1 \cdot 0 = \$(\text{US 1968})\ 0 \cdot 65$$

and

$$\$(\text{US 1968})\ 2 \cdot 4 = \pounds(1968)\ 1 \cdot 0$$

Also Casper *et al.*[72] give the energy intensity of the outputs of a number of heavy engineering sectors as 220 MJ(th) per £ (1968). Thus if it is assumed that the plant is on-stream for 90% of its 20-year lifetime, and 50% of initial costs are allowed for maintenance, then the energy requirement of plant and equipment

$$= 1·5 \times 533 \times 10^6 \times \frac{0·65}{2·4} \times 220 \times \frac{1}{20 \times 365 \times 0·9}$$
$$= 7·25 \times 10^6 \text{ MJ(th) per standard day}$$

Catalysts and chemicals
Estimated annual cost of catalysts and chemicals was $ (US 1975) 8 million. Again using ref. 71 for 'chemicals and allied products',

$$\$ (\text{US } 1975) \ 1·0 = \$ (\text{US } 1968) \ 0·55$$

and

$$\$ (\text{US } 1968) \ 2·4 = £ (1968) \ 1·0$$

And using energy intensity, as given in ref. 72 for 'general chemicals' as 702 MJ(th) per £ (1968), and assuming that plant is on-stream 90% of the time, the energy requirement of catalysts and chemicals

$$= 8·0 \times 10^6 \times \frac{0·55}{2·4} \times 702 \times \frac{1}{365 \times 0·9}$$
$$= 3·918 \times 10^6 \text{ MJ(th) per standard day}$$

total GER of process $= 652·2 \times 10^6$ MJ per standard day

Outputs
The Sasol process gives a wide variety of products and these are tabulated in Table 11.

% of total plant energy requirements apportioned to motor spirit production

$$= \frac{134·50}{267·2}$$
$$= 46·91\%$$

Thermal efficiency of process
(a) with respect to motor spirit only $= 20·6\%$
(b) with respect to all products $= 41·0\%$

METHANOL PRODUCTION

Methanol can be produced directly from coal by gasification followed by reaction of the resulting carbon monoxide and hydrogen. Other byproduct gases are scrubbed off. Three such coal gasification processes have been proved to be commercial—the Lurgi, the Koppers/Totzek and the Winkler processes. The Hygas-Electrothermal reactor is also another successful variant of the gasification process. All these gasification reactions require high temperatures (900–1400 °C). Coal is treated with high pressure steam and compressed air, and the products usually are in the range of methanol (a product of synthesis gas reaction), ammonia, naphtha, phenol and tars.

In an earlier work, Edewor[74] showed that to make 200 short tons of methanol per day from coal using the Fischer–Tropsch technique would require 320 tonnes of coal per day (energy involvements calculated as 5·0–10·0 billion Btu per day). This should give an average GER equivalent to $7·90 \times 10^6$ MJ per day for 200 short tons per day. In terms of GER per tonne of methanol produced this would equal

$$\frac{7·90 \times 10^6}{200} \times \frac{2000}{2240} = 35·3 \times 10^3 \text{ MJ per tonne}$$

Also, in another variant of Edewor's approach,[74] the estimated energy involvements for methanol synthesis based on underground gasification of coal yielded figures of 10·0–18·00 billion Btu for a production of 200 short tons of methanol and using up of 565 tonnes of coal per day. In terms of GER per tonne of methanol this averaged $66·0 \times 10^3$ MJ(th) per tonne of methanol produced.

In 1968, ICI developed a low pressure process, using 50 atm at 250 °C and a highly selective copper-based catalyst, to synthesise methanol from the product gases of the gasification stage.

Chan[19] prepared a conceptual design of a plant producing methanol from coal via synthesis gas as in the Sasol plant in South Africa. The following calculations are based on his design.

Inputs

Coal
For a plant producing 10 121 tonnes of methanol per standard day, Chan[19] estimated the total coal input as 28 904 tonnes per day (CV 20 600 MJ(th) per tonne). Therefore

total energy input from coal $= 28\,904 \times 26\,604$
$$= 594 \cdot 68 \times 10^6 \text{ MJ(th) per standard day}$$

Plant and capital equipment
Estimated cost of plant was $(US 1975) 472 million. As in the previous treatment of refs. 71 and 72

$$\$(\text{US } 1975)\ 1 \cdot 0 = \$(\text{US } 1968)\ 0 \cdot 65$$

and

$$\$(\text{US } 1968)\ 2 \cdot 4 = £(1968)\ 1 \cdot 0$$

and energy intensity of capital

$$= 220 \text{ MJ(th) per } £(1968)$$

Thus, if the plant is on-stream for 90% of its 20-year lifetime and 50% of initial costs are allowed for maintenance, the energy requirement of plant and equipment

$$= 1 \cdot 5 \times 472 \times 10^6 \times \frac{0 \cdot 65}{2 \cdot 4} \times 220 \times \frac{1}{20 \times 365 \times 0 \cdot 9}$$
$$= 6 \cdot 42 \times 10^6 \text{ MJ(th) per standard day}$$

Catalyst and chemicals
Chan[19] estimated the annual cost of catalyst and chemicals as $(US 1975) 6 million. Using the method shown above, energy requirements of catalyst and chemicals

$$= 2 \cdot 94 \times 10^6 \text{ MJ(th) per standard day}$$

Therefore

$$\text{GER of process} = (594 \cdot 68 + 6 \cdot 42 + 2 \cdot 94) \times 10^6$$
$$= 604 \cdot 04 \times 10^6 \text{ MJ(th) per standard day}$$

Outputs
Again, this process yields several products and these are shown in Table 12.
% of total energy requirement for plant apportioned to methanol

$$= \frac{233 \cdot 61}{336 \cdot 79}$$
$$= 69 \cdot 36\%$$

Therefore

$$\text{GER of methanol} = \frac{1}{10\ 121} \times 0{\cdot}6936 \times 604{\cdot}04 \times 10^6$$
$$= 41\ 398\ \text{MJ per tonne of methanol}$$

and

$$\text{NER of methanol} = 41\ 398 - 23\ 100$$
$$= 18\ 298\ \text{MJ per tonne}$$

Using an average figure of 7·5 billion Btu,[74] a GER value of 35 300 MJ per tonne is obtained.

The thermal efficiency of the process
(a) with respect to methanol only $= 38{\cdot}7\%$
(b) with respect to all products $= 55{\cdot}8\%$

Hydrogenative Extraction of Coal with Anthracene Oils

INTRODUCTION

This is a typical CSF-type process being investigated by the National Coal Board (NCB) at the research centre at Cheltenham. The process involves the use of anthracene oils (coal-derived solvent) to extract the volatile components of coal at about 400–480 °C and partial pressure of hydrogen of about 200–340 bars. According to Davies et al.,[20]a 60% yield of aromatic distillates makes them the major components. The conventional catalyst used is cobalt–molybdenum but investigations into other catalysts are under way. It is estimated that up to 85% of the coal substance is dissolved by the solvent.[20]

In a demonstration design, a catalyst–coal feedstock ratio of about 1:4 to 1:20 is used; a hydrogen–carbon ratio of 0·62 is obtainable in the extract before hydrogenation and a value of 0·90 is obtainable after the hydrogenation stage. Plant investment was £(1974)42 million for a plant throughput of 1 million tonnes of coal per annum. Product output of about 45% of coal feed is obtainable. The main products obtained are gasoline, gas oil and hydrocracking residue which is converted to electrode coke. An overall thermal efficiency of 60% for the process was calculated and the cost of products per tonne was given as £(1974)50·0.

EVALUATION OF GER OF PROCESS

Inputs

Coal
'Energy Resources'[69] gives a range of calorific values for British bituminous – sub-bituminous coals from 7·9 to 10 kWh per kilogram; the CV of coal feed is calculated as 32 950 MJ(th) per tonne. Davies *et al.*[20] designed a process to handle 1 million tonnes of coal per year. Therefore

daily feed of coal   = 2740 tonnes
moisture content of coal = 16%

Hence energy input from coal   = 2740 × 32 950
                                = 90·28 × 10^6 MJ(th) per standard day

Electricity
From calculations by the Consolidation Coal Company[16] it can be estimated that about 79·3 tonnes of extra coal per standard day will be required for electricity purposes. Therefore

estimated extra energy input = 79·3 × 32 950
                             = 2·61 × 10^6 MJ(th) per standard day

Plant and capital equipment
This is estimated at £(1974)42 million. At 1978 price levels the estimated cost will be £(1978)61·50 million and at 1968 price levels this is equivalent to £28·7 million. From Casper *et al.*,[72] the energy intensity of capital equipment = 220 MJ(th) per £(1968). Thus, assuming plant to be on-stream for 90% of its 20-year lifetime and allowing about 50% of initial costs for maintenance, the energy requirement of the plant is

$$= 1·5 × 28·7 × 10^6 × 220 × \frac{1}{20 × 365 × 0·9}$$
$$= 1·44 × 10^6 \text{ MJ(th) per standard day}$$

Catalyst and chemicals
Davies *et al.*[20] estimate that hydrogenation costs account for about 41% of total plant investment. Suppose catalyst cost accounts for 25% of the hydrogenation costs, then the costs of catalysts and related chemicals can

be estimated to be £(1974)4·30 million, or £(1968)2·93 million. From Casper *et al.*,[72] the energy intensity of the catalyst and general chemicals = 702 MJ(th) per £(1968). Thus

$$\text{energy requirement of catalyst, etc.} = 2\text{·}93 \times 10^6 \times 702 \times \frac{1}{365 \times 0\text{·}9}$$

$$= 6\text{·}4 \times 10^6 \text{ MJ(th) per standard day}$$

Preheat stage: An estimated additional 201 tonnes of coal per standard day are required to give the preheat required for the initial coal feed, i.e. another energy requirement of

$$6\text{·}63 \times 10^6 \text{ MJ(th) per standard day}$$

Hydrogen and oxygen production
In the original design of this process, natural gas is reformed to provide hydrogen for the hydrogenation stage. However, in this analysis it will be assumed that coal will be reformed to give the required hydrogen.

H–C ratio of solvent extract   = 0·62
H–C ratio of hydrogenated filtrate = 0·90
hydrogen content of coal extract   = 4·7%

Therefore for an extract production rate of 1205·6 tonnes per day, *in situ* hydrogen present in extract = 56·7 tonnes per standard day. The extra hydrogen required to raise the H – C ratio from 0·62 to 0·90 is calculated as 25·6 tonnes per standard day.

$$\frac{1}{n}CH_{0\text{·}8n}\,(s) + 2H_2O(l) \longrightarrow 2\text{·}40\,H_2(g) + CO_2(g) \qquad (12)$$

From the coal reforming equation given above, it can be calculated that 1 tonne of hydrogen requires 2·67 tonnes of moisture ash free (MAF) coal. Also

$$\frac{1}{n}CH_{0\text{·}8n}\,(s) + 1\text{·}20\,O_2(g) \longrightarrow CO_2(g) + 0\text{·}40\,H_2O(l)$$

$$\Delta H = -97\text{·}0 \text{ kcal} \qquad (13)$$

$$H_2(g) + \tfrac{1}{2}O_2(g) \longrightarrow H_2O(l)\ \Delta H = -68\text{·}3 \text{ kcal} \qquad (14)$$

By addition, eqn. (13) – 2·40 eqn. (14) gives eqn. (12), $\Delta H = -70\text{·}3$ kcal. Thus each tonne of coal used in hydrogen production requires a theoretical minimum of 22 801 MJ(th) or 0·725 tonnes of MAF coal equivalent to sustain the reaction. But coal as received = 16% moisture by weight. Therefore 1 tonne of hydrogen requires 4·60 tonnes of MAF coal or

$$4.60 \times \frac{116}{100} = 5·34 \text{ tonnes of coal}$$

From the above equation it can also be shown that 1 tonne of hydrogen requires about 5·88 tonnes of oxygen. Harris[73] gives the replacement energy requirement of oxygen as 6·33 GJ per tonne, i.e. to produce 1 tonne of oxygen will need about 6·33 GJ.

Thus input energy requirements to
(a) hydrogen production = 25·6 × 5·34 × 32 950
$$= 4·50 \times 10^6 \text{MJ(th) per standard day}$$
(b) oxygen production   = 6330 × 5·88 × 25·6
$$= 0·953 \times 10^6 \text{ MJ(th) per standard day}$$
Finally

$$
\begin{aligned}
\text{total GER of process} &= (90·28 + 2·61 + 1·44 + 6·4 + 6·63 \\
&\quad + 4·50 + 0·953) \times 10^6 \\
&= 113·2 \times 10^6 \text{ MJ(th) per standard day}
\end{aligned}
$$

Outputs
The main products are

gasoline (13% of coal feed) = 356·2 tonnes per standard day (CV = 48 671 MJ(th) per tonne)
gas oil (27% of coal feed) = 739·8 tonnes per standard day (CV = 44 091 MJ(th) per tonne)
electrode coke (4% of coal feed) = 109·6 tonnes per standard day (CV = 19 520 MJ(th) per tonne)
  other side products, total energy = 15·8 × 10^6 MJ(th) per standard day

% total plant requirement apportioned to main products

$$
\begin{aligned}
&= \frac{52·1 \times 10^6}{67·9 \times 10^6} \\
&= 76·73\%
\end{aligned}
$$

Thermal efficiency of process
(a) with respect to main products $= \dfrac{52·1 \times 10^6}{113·2 \times 10^6}$
$$= 46\%$$
(b) with respect to all products = 60%

Supercritical Extraction of Coal with Toluene

INTRODUCTION

This process is based on the fact that most volatile substances are able to vaporise more freely in the presence of a compressed gas. It is well known that the solvent power of a gas increases with its density and the density is greatest, for any given applied pressure, at the critical temperature of the gas.

Thus in this process the solvent chosen, toluene, is vaporised and heated to about its critical temperature (400 °C). The coal to be treated is also heated to this temperature and contacted with the vaporised solvent. Volatile materials in the coal pass into the gaseous phase and are carried along by the toluene carrier. This is done in the extractor,[21] and the residual part, mainly char, is steam purged and stored. The resulting extract – toluene mixture undergoes pressure reduction in a flash column where the volatile materials of the coal are redeposited. The toluene is cleaned and recycled again, and the resulting coal extract is cleaned and collected as main product.

In this process no hydrogen is required. About one-third of the total coal material could be extracted by this method and a typical coal feed is about 5·2% moisture content. Gross heating value of coal (CV) is approximately 34 131 MJ(th) per tonne while that of the extract is about 36 726 MJ per tonne. As of 1974 the total investment of plant was $ 140·1 million for an annual throughput of 3·65 million tonnes of coal. It was estimated by

Maddocks and Gibson[21] that the cost of producing the coal extract by this method was $ 9·9 per barrel or £ (1974)38·5 per tonne. The extract is a low-melting glassy solid with a ring and ball index of about 70 °C. It has a generally open-chain polynuclear aromatic structure linked by ether and methylene groups. To be of any use as a feedstock to refineries this extract has to be further hydrogenated.

EVALUATION OF GER OF PROCESS

Inputs

Coal

Expected throughput of coal[21] = 10 968 tonnes per standard day

H – C aromatic ratio of coal   = 0·72

Moisture content of coal   = 5·2%

Calorific value of coal   = 34 131 MJ per tonne

Heat input from coal   = 34131 × 10 968

         = 3·74 × 10⁸ MJ per day

Plant is operated for 330 days per year

Electricity

Estimated electricity required (from ref. 16) = 406 tonnes equivalent of coal per standard day. Energy equivalent of this = 0·139 × 10⁸MJ per standard day.

Plant and equipment

Estimated total investment[21] = $ (1974) 140·1 million

                       = £ (1974) 77·83 million

        £ (1974) 1·0 = $ (1974) 1·80

Also $ (1974) 1·0 = $ (1968) 0·74 (ref. 71)

and   £ (1968) 1·0 = $ (1968) 2·4

According to Casper *et al.*,[72] the energy intensity of heavy engineering equipment for the UK = 220 MJ per £ (1968). Therefore

energy requirement (in terms of plant contribution)

$$= 1·5 \times 140·1 \times 10^6 \times \frac{0·74}{2·4} \times \frac{220}{20 \times 330}$$

$$= 0·022 \times 10^8 \text{ MJ per standard day}$$

Note: The above equation assumes a 20-year lifetime for the plant, 330 working days in a year and 50% of capital investment set aside for maintenance parts for the plant.

Chemicals and operating supplies
Annual costs of these are put at $ (1974) 2·3 million. According to ref. 71 for 'general chemicals'

$$\$ (1974)\ 1{\cdot}0 = \$ (1968)\ 0{\cdot}68$$
$$\$ (1968)\ 2{\cdot}4 = £ (1968)\ 1{\cdot}0$$

Also from Casper *et al.*,[72]

energy intensity of general chemicals $= 702\ \text{MJ per } £ (1968)$

Therefore

$$\text{energy requirements} = 2{\cdot}3 \times 10^6 \times \frac{0{\cdot}68}{2{\cdot}4} \times \frac{702}{330}$$
$$= 0{\cdot}014 \times 10^8\ \text{MJ per standard day}$$

Make-up toluene
In this process a provision is made for losses in solvent. Make-up toluene is estimated at 1·2 tonnes per hour, or 28·28 tonnes per standard day. Cost of make-up toluene is put at $ 550 per gallon, i.e. $ 164 per tonne, taking 1 gallon = 7·5 lb. Therefore cost of make-up toluene is $ (1974) 4·64 × 10³ per standard day.
Applying the same method as above

$$\text{energy requirement} = 4{\cdot}64 \times 10^3 \times \frac{0{\cdot}68}{2{\cdot}4} \times 702$$
$$= 0{\cdot}92 \times 10^6\ \text{MJ per standard day}$$

Therefore

total GER $= 3{\cdot}92 \times 10^8\ \text{MJ per standard day}$

Outputs
The main product, the extract, is estimated at 2983·2 tonnes per standard day; the CV of this is 36 726 MJ per tonne. Residue, non-caking porous solid with an appreciable volatile material, ideal for gasification, is also a product with a production rate of 5680·8 tonnes per standard day, and the CV of the residue is 33 092 MJ per tonne.

% of total plant energy requirement apportionable to extract production

$$= \frac{1 \cdot 096 \times 10^8}{2 \cdot 976 \times 10^8}$$

$$= 36 \cdot 83\%$$

Thermal efficiency of process

(a) with respect to extract only $= 28\%$
(b) with respect to extract of residue $= 75 \cdot 73\%$

APPENDIX B1

Energy Requirements for a Plant Processing 36 tonnes per day of Municipal Refuse to Oil

Energy inputs

Electricity

Kaufman and Weiss[25] put the daily electrical requirements of such a plant at 4288 kWh per day. Assuming that the generation of electricity is 30% efficient this will give a requirement, in thermal terms, of

$$4288 \times 3 \cdot 6 \times 3 \cdot 3$$
$$= 50 \cdot 94 \times 10^3 \text{ MJ(th) per day}$$

Plant and capital equipment

Total plant costs are estimated at $(1975) 1 116 906 by Kaufman and Weiss.[25] At 1978 price levels (10% inflation assumed each year)

$$\text{equivalent investment} = \$(1978)\ 1\ 486\ 602$$
$$= \$(1968)\quad 573\ 150$$

But from ref. 71 $(1968) 2·4 = £(1968) 1·0. Also, from Casper *et al.*,[72] the energy intensity of heavy engineering products is about 220 MJ(th) per £(1968). Kaufman and Weiss estimate a peak period of about 15 years of operation for such a plant. Thus assuming a 20-year lifetime for the plant and also allowing 50% of the initial costs for plant maintenance

$$\text{energy requirement of equipment} = \frac{1 \cdot 5 \times 573\ 150}{2 \cdot 4} \times \frac{220}{20 \times 260}$$
$$= 15 \cdot 16 \times 10^3 \text{ MJ(th) per day}$$

Catalysts and chemicals
Annual costs of nickel catalyst and activated carbon used in plant are given
as $(1975) 23 000. At 1968 price levels these could be put as $(1968)
11 803. From ref. 71 $(1968) 2·40 = £(1968) 1·0. From ref. 72 the energy
intensity of 'general chemicals' is 702 MJ(th) per £(1968). Therefore

$$\text{energy requirement of chemicals and catalysts} = \frac{11\ 803}{2\cdot4} \times \frac{702}{260}$$

$$= 13\cdot38 \times 10^3 \text{ MJ(th) per day}$$

Hydrogen production
In the original design, natural gas was used as a feedstock to the steam
reforming plant. But Kaufman and Weiss indicate that the light hydrocarbon fraction of the product could be recycled and reformed instead of
natural gas. This makes the plant self-sustaining. Light oil demand of the
hydrogen plant is estimated at 0·10 barrels per hour;[25] the CV of the oil is
put at about 43 × 10³ MJ(th) per tonne, and thus the energy requirement of
the hydrogen plant is given by

$$\frac{0\cdot10 \times 24 \times 43 \times 10^3}{7\cdot0} \quad (7 \text{ barrels} = 1 \text{ tonne})$$

$$= 14\cdot74 \times 10^3 \text{ MJ per dy}$$

Boiler
The boiler raises steam and uses natural gas in the original design. But 0·14
barrels per hour of byproduct light hydrocarbons of the process can be
used instead.[25] Hence

$$\text{energy requirement of boiler} = 0\cdot14 \times 24 \times \frac{43 \times 10^3}{7}$$

$$= 20\cdot64 \times 10^3 \text{ MJ per day}$$

Furnace
The feed to the reactor (ground refuse, catalyst and solvent oil) is heated by
the furnace to about 455 °C (850 °F) before reaction occurs in the reactor.
The furnace uses natural gas, but again the light hydrocarbon product
could be used. According to Kaufman and Weiss about 0·42 barrels per
hour of this hydrocarbon is needed,[25] i.e.

$$\text{energy requirement} = 0\cdot42 \times 24 \times \frac{43 \times 10^3}{7}$$

$$= 61\cdot43 \times 10^3 \text{ MJ per day}$$

Refuse

Daily throughput of refuse is given by Kaufman and Weiss as 36 tonnes of combustible refuse per day, and since the estimated calorific value of the refuse is 5000 calories per gram or $21\cdot3 \times 10^3$ MJ per tonne,

$$\text{energy input from refuse} = 21\cdot3 \times 10^3 \times 36$$
$$= 766\cdot8 \times 10^3 \text{ MJ per day}$$

Refuse collection

This is an important aspect of the process. In the Holden case design, about 50% of the weekly refuse arrives at the plant during the weekend. Collection can be made easy with the help of the population. In many developed cities the municipal waste disposal departments accomplish the collection with trucks of capacities 20–50 cubic yards. Because of the general non-availability of published information on this, the data of Bidwell and Mason[34] on the Greater London area will be used. The Greater Manchester area follows similar patterns. Reference 34 gives the cost of transportation of refuse from the Greater London area to the refuse plant as 4 pence per tonne-kilometre for 1975 (one-way movement only). Also, the cost of transportation of refuse-derived oil to the consumer is estimated to be another 4 pence per tonne-kilometre. Suppose the plant location is about 30 km from the town population, and 6·0 pence per tonne-kilometre is assumed for 1978 price level costs of refuse transportation, then the cost of collecting refuse and transporting to the plant site (25% non-combustible in town refuse) will be

$$6\cdot0 \times 48 \times 30 = \pounds(1978)\ 86\cdot40 \text{ per day}$$

Energy Requirements of Plants with Capacities Greater than 36 tonnes per day

In calculating the energy requirements of such plants, scale-up procedures are required. For the furnaces, boilers, etc., a direct linear scale-up is required, i.e. factors of the requirements for the 36 tonnes plant will be used. However, in plant and capital equipment the six-tenths rule will be used, i.e.

$$\left[\frac{(\text{capacity})_2}{(\text{capacity})_1} \right]^{0.6} = \frac{(\text{cost})_2}{(\text{cost})_1}$$

which implies that the $(\text{cost})_2$ for new plant, with $(\text{capacity})_2$, is given by

$$(\text{cost})_1 \times \left[\frac{(\text{capacity})_2}{(\text{capacity})_1} \right]^{0.6}$$

Then using the same energy–cost conversion procedures as in Appendix B1, the energy requirements of plants with higher capacities can be calculated. These are for capacities of 100 tonnes per day, 500 tonnes per day, 1000 tonnes per day and 2000 tonnes per day. Table 24 gives the summary of calculations. The X_f values vary slightly because about 87% of the total energy input comes from the refuse heat value.

APPENDIX B3

Energy Requirements of Specially Grown Crops for Oil and Gas Production

SUPPORT ENERGY

This is the energy input required to make the crops grow at faster rates, and includes energy for producing fertilisers, fuels for tractors, herbicides and other field operations. In 1973 the total agricultural use of such support energy in the UK was equivalent to 20% of the energy fixed by crop photosynthesis.[35] In that year the total agricultural use of support energy was put at 246×10^9 MJ. Grass, sugar cane and maize are potential energy carrier crops for the future. Sample analyses were carried out by Walsingham and Spedding on perennial rye grass and lucerne, and Table 35 shows the results as presented by Cooper.[35] From the table it can be seen that to produce 1 tonne of rye grass will require approximately $3 \cdot 57 \times 10^3$ MJ of energy per year while lucerne will need about 985 MJ per year. Leach[37] estimates the energy input to production of phosphate fertiliser as 2260 MJ per tonne, that of potash fertiliser as 5600 MJ per tonne and that of nitrogenous fertiliser as 37 526 MJ per tonne. It can be seen that fractions of total energy requirements of individual fertilisers can produce as much as $10-12$ times the crop output. The energy requirements for growth and collection of crops such as maize (corn) and sugar cane or sugar beet vary with the locality of growth and the types of labour and fertilisers used. From the works of Hill[63] and Pimentel et al.,[75] Table 36, the support energies of the various crops (corn and sugar), can be constructed. Thus, mean values of support energies for crop growth could be expressed as

grass	3850 MJ per tonne-year of raw feed
maize	5250 MJ per tonne-year of raw feed
sugar beet	2733 MJ per tonne-year of raw feed
corn (mechanical farming)	3200 MJ per tonne-year of raw feed

However, only the case for grass as feedstock is considered here.

Collection (Gathering) Energy

Most food materials travel by road in the UK. Thus the most likely mode of collection of these specially grown crops, in this case dry grass or hay, will be by road. Leach[37] presents a compilation of fuel energy inputs for some vehicles. The results, shown in Table 37, are based on the data provided by the UK Transport and Road Research Laboratory in Berkshire. The Leyland 12 tonne truck will be an ideal vehicle for collection of these crops and crop wastes. Locality of use will be rural and most of the time the lorries will have a full load. With these specifications fuel energy input for a lorry-load of grown crops will be about 12 MJ(th) per vehicle-kilometre. The weight of this type of lorry is given as 7·45 tonnes when unladen and the energy contribution of this weight of steel has to be taken into account. Suppose the truck material is assumed to be 95% steel and 5% copper, having energies of fabrication of 46 700 MJ per tonne and 140 000 MJ per tonne respectively. Then a weight of 7·45 tonnes should have a total energy value of 383×10^3 MJ. With an assumed lifetime of a lorry of 6 years, with a working schedule of 12 hours per day and 260 working days per annum, the amortised energy contribution of such a lorry per day will be

$$\frac{383 \times 10^3}{6 \times 260} = 245 \text{ MJ per day}$$

From Appendix B5 it is estimated that 11 lorries will be required to collect about 1000 tonnes of feed materials per day. This implies an energy requirement of

$$245 \times 11 = 0.003 \times 10^6 \text{ MJ per day}$$

Energy Requirements for Producing Oil from Specially Grown Crops

Electricity

Kaufman and Weiss[25] indicate that a plant processing 36 tonnes of combustible municipal refuse should need about 4287·7 kWh per day of electrical energy. This includes power for pumps, hammer mill and other electrically driven equipment. From this, an energy requirement for a plant of capacity 1000 tonnes per day can be estimated at 119 103 kWh per day. But 1 kWh(e) = 3·6 MJ(e) and if 30% efficiency is assumed in electricity generation the energy requirement of the plant will be

$$119\ 103 \times 3·6 \times 3·3 = 1·42 \times 10^6 \text{ MJ(th) per day}$$

Plant and capital equipment

Following the six-tenths rule of capacity – cost estimates, the cost of a plant processing 1000 tonnes of grown crops could be put at $(1978)10·92 million or $(1968) 4·21 million, on the basis of 10% inflation each year. But from ref. 71

$$\$(1968)2·40 = £(1968)1·0$$

Also from Casper et al.,[72] the energy intensity of heavy engineering products

$$= 220 \text{ MJ(th) per £(1968)}$$

Assuming another 50% of the initial costs are set aside for maintenance purposes and the plant is expected to have a 20-year working life of 260 days per year, then the energy requirement of this input is

$$\frac{1 \cdot 5 \times 4 \cdot 21 \times 10^6 \times 220}{2 \cdot 40} \times \frac{1}{20} \times \frac{1}{260}$$

$$= 0 \cdot 110 \times 10^6 \, \text{MJ(th) per day}$$

Catalysts and chemicals

These are treated in a manner similar to that of the Holden Town design. The annual cost of activated carbon and nickel catalysts is put at $(1975)169\,023$ or $(1968)86\,735$. But, again, from ref. 71

$$\$(1968)2 \cdot 4 = \pounds(1968)1 \cdot 0$$

and from Casper *et al.*,[72] the energy intensity of allied chemicals is

$$702 \, \text{MJ(th) per } \pounds(1968)$$

Therefore the energy requirement by way of catalyst and chemical input

$$= \frac{86\,735}{2 \cdot 4} \times \frac{702}{1} \times \frac{1}{260}$$

$$= 0 \cdot 098 \times 10^6 \, \text{MJ(th) per day}$$

Furnace and boiler feed-gas

In a calculation similar to Appendix B1 the estimated energy requirement is put at 15·56 barrels per hour of oil equivalent (light hydrocarbons from the process plant are recycled and used). This gives an energy requirement, in terms of thermal joules, of

$$15 \cdot 56 \times 24 \times \frac{43 \times 10^3}{7} = 2 \cdot 3 \times 10^6 \, \text{MJ(th) per day}$$

where the heating value of product hydrocarbons is 43×10^3 MJ(th) per tonne.

Hydrogen production

Part of the product oil is used also to produce hydrogen in the reforming unit. It is estimated that 2·78 barrels per hour of oil will be needed, i.e. the energy requirement will be

$$2 \cdot 78 \times 24 \times \frac{43 \times 10^3}{7} = 0 \cdot 41 \times 10^6 \, \text{MJ(th) per day}$$

Support energy

It has been shown that the support energy for growth of grass is about 3850 MJ(th) per tonne-year (Table 36) or

$$\frac{3850}{330} = 11 \cdot 67 \, \text{MJ(th) per tonne-day}$$

for a growth period of 330 days in a year. Therefore for 1000 tonnes per day of crops, the energy requirement by way of support energy

$$= 11 \cdot 67 \times 1000$$
$$= 0 \cdot 012 \times 10^6 \, \text{MJ(th) per day}$$

Collection and transport
Distance between farmland and plant site is assumed to be 30 kilometres. Also the 12 tonne Leyland vehicle has been assumed to be the mode of transporting the grown crop. At an average speed of trucks of 48 kilometres per hour a truck will make a return trip in 1·25 hours. Allowing 20 min for loading and unloading the vehicle each lorry will make a complete return trip in 1·60 hours. Thus for a day operation of 12 hours a lorry can make about seven trips. Also, from truck specifications,

weight of truck when unladen = 7·45 tonnes
weight of truck with full load = 13·03 tonnes

Therefore the estimated load capacity = 13·0 tonnes, i.e. for seven trips each truck can carry

$$13 \cdot 0 \times 7 \, \text{tonnes} = 91 \cdot 0 \, \text{tonnes of crop materials}$$

These should be in the form of bales of dried grass or hay. For 1000 tonnes per day of crop requirement about 1000/91 or 11 lorries will be needed to do the job. Hence, energy input from transportation will be:

$$\text{for vehicle trips} = 11 \times 12 \times 30$$
$$= 0 \cdot 004 \times 10^6 \, \text{MJ(th) per day}$$

$$\text{from vehicle parts} = 245 \times 11$$
$$= 0 \cdot 003 \times 10^6 \, \text{MJ per day}$$

Total energy needed for collection and transportation of crop material is thus about $0 \cdot 007 \times 10^6$ MJ per day.

Leach[37] presents the hourly rates of a 74 hp 55 kW tractor working on a grass and hay farm as approximately 10·8 hours per hectare. A standard 50 hp 37 kW tractor should average about 12·6 hours per hectare. From Table 34 the growth rate of perennial rye grass is about 25 tonnes dry matter yield per hectare-year; theoretically this species of grass grows for

365 days in a year. However, about 330 days will be used to calculate land requirements. A crop material need of 1000 tonnes per day will require

$$\frac{1000 \times 330}{25}$$

or 13 200 hectares of land for grass cultivation.

A tractor rate of 12·6 hours per hectare will mean a total of 166 320 tractor hours per year. Leach[37] also states that a standard 50 hp tractor has a working life of 6000 hours or 900 hours per annum. Thus for every 6·67 years about 185 tractors are required for the job of cultivation and collection of rye grass. Energy requirements of these are included in the support energy for crop growth. Hence, energy requirements in this section are those due to trucks.

APPENDIX B6

Energy Requirements of San Diego Plant

Electricity
Levy[39] put the plant electrical requirement at 136 kWh per short ton of refuse processed. This is used to run the various equipment. At a conversion rate of 1 kWh $= 3.6$ MJ(e) and electricity generation at 30% efficiency, this is equivalent to a daily need of

$$200 \times 136 \times 3.6 \times 3.3 = 0.323 \times 10^6 \text{ MJ(th) per day}$$

Plant and capital equipment
Levy[39] estimates the cost of plant equipment as $(1974) 6·3 million. At 1978 price levels, assuming 10% inflation each year, this will be equivalent to $(1978) 9·22 million or £(1978) 4·7 million, assuming £(1978) 1·0 is equivalent to $(1978) 1·95. From Casper et al.,[72] the energy intensity of heavy engineering products is about 220 MJ(th) per £(1968). Suppose plant costs are reduced to £(1968) levels and assuming another 50% of the initial plant costs are set aside for maintenance purposes over a 20-year lifespan of plant, then energy contributions of plant equipment are

$$\frac{1.5 \times 1.8 \times 10^6 \times 220}{20 \times 330} = 0.091 \times 10^6 \text{ MJ(th) per day}$$

where 1 year is taken as 330 operating days.

Refuse material
The heat value of input refuse is about 9.70×10^3 MJ per tonne,[39] and with a capacity of 200 tonnes per day this is equivalent to a daily energy input of

$$9.7 \times 10^3 \times 200 = 1.94 \times 10^6 \text{ MJ(th) per day}$$

Heat for burners, etc.
Levy[39] estimates that 2·5 litres of product oil equivalent will be required to provide the process heat for each tonne of refuse processed. At an average heating value of 32 MJ(th) per litre, the energy contributions are

$$200 \times 2.5 \times 32 = 0.016 \times 10^6 \text{ MJ(th) per day}$$

Collection of refuse
This part of the process could be dealt with either in terms of collection labour costs, as in the Holden case study, or analysed into sub-energy inputs, as in the case of specially grown crops. But because refuse collection could fluctuate in pattern with locality the better approach will be in terms of labour costs.

Suppose, as in the Holden case study, a transportation cost of 6·0 pence per tonne-kilometre is assumed, and suppose the pyrolysis plant is located 30 kilometres from the city population. Then the cost of transporting refuse to the plant is estimated as

$$30 \times 200 \times 6 = £\,360 \text{ per day}$$

APPENDIX B7

Energy Requirements of Plants with Capacities Greater than 200 tonnes per day

Based on 2000 tonnes of refuse throughput per day the following are the energy requirements of a typical Garret Plant.

Electricity
On a 136 kWh per short ton requirement the electrical energy input per day is equivalent to

$$2000 \times 136 \times 3.6 \times 3.3 = 3.23 \times 10^6 \text{ MJ(th) per day}$$

Plant and capital equipment
Based on the six-tenths rule of cost – capacity estimate, the plant cost for a 2000 tonne capacity is given by

$$\left(\frac{2000}{200}\right)^{0.6} = \frac{C_2}{1.8 \times 10^6}$$

where C_2 is the cost of a 2000 tonne per day plant and 1.8×10^6 is the cost of a 200 tonne per day plant in £ (1968). Therefore

$$C_2 = £(1968)\ 7.21 \text{ million}$$

Following the procedure in Appendix B6, the energy requirement by way of plant equipment

$$= \frac{1.5 \times 7.21 \times 10^6 \times 220}{20 \times 330}$$

$$= 0.361 \times 10^6 \text{ MJ(th) per day}$$

215

Refuse material
As in Appendix B6, the energy input by way of refuse material is given by

$$9 \cdot 7 \times 10^3 \times 2000 = 19 \cdot 4 \times 10^6 \text{ MJ(th) per day}$$

Heat for burners, etc.
Again, using Levy's[39] estimate of 2·5 litres of oil per tonne of processed refuse as the energy requirement (Appendix B6), the daily input for heat purposes

$$= 2000 \times 2 \cdot 5 \times 32$$
$$= 0 \cdot 16 \times 10^6 \text{ MJ(th) per day}$$

Collection of refuse
The cost per tonne of refuse will still be taken as that in the case of 200 tonnes per day capacity, i.e. 6 pence per tonne-kilometre. Therefore

$$\text{cost of collection} = 2000 \times 30 \times 6$$
$$= £ 3600 \text{ per day}$$

The total GER of the process is equal to $23 \cdot 15 \times 10^6$ MJ(th) per day. The estimated outputs of products are in the same ratio as in the case of 200 tonnes per day capacity, i.e. 2000 tonnes of refuse giving 723 tonnes of synfuels (oil, gas and char). A summary of GER, NER and X_f values of both 2000 and 1000 tonnes capacities is given in Table 40.

APPENDIX B8

Energy Requirements in Bioconversion Processes: Leafy Material Digestion

Inputs

Plants and capital equipment

Sitton and Gaddy[29] give the capital investments of plant as $(1974)45·7 million including 30% contingency. With this level of investment of $46 million, including the added contingency, and allowing for an inflationary trend of 10% each year, the estimated investment level as of 1978 will be

$$\$46 \times 10^6 = (1·1)^4$$
$$= \$(1978)67·35 \text{ million}$$
$$= \$(1968)26·0 \quad \text{million}$$

From ref. 71

$$\$(1968)2·4 = £(1968)1·0$$

Also, from Casper *et al.*,[72] the energy intensity of heavy engineering products

$$= 220 \text{ MJ(th) per } £(1968)$$

Though no indication of lifetime of the plant was made by the authors, an assumed period of 20 years will not be too far wrong. Also, suppose about 50% of initial plant cost is set aside for maintenance materials, then the energy requirements contributed by the plant will be those amortised over the 20-year lifetime. Therefore, assuming an operational period of 330 days per year,

217

$$\text{energy requirement of plant equipment} = \frac{1\cdot5 \times 26\cdot0 \times 10^6 \times 220}{2\cdot4 \times 20 \times 330}$$
$$= 0\cdot54 \times 10^6 \text{ MJ(th) per day}$$

Crop material feed
In the design, 4380 tonnes of biomass is required daily. The average heating value of crop material is $15\cdot4 \times 10^3$ MJ(th) per tonne.[29] Therefore

$$\text{daily energy input by way of crop material} = 15\cdot4 \times 10^3 \times 4380$$
$$= 67\cdot3 \times 10^6 \text{ MJ(th) per day}$$

Power
(a) Grinding. Hammer mill/shredder power requirements are estimated by FEA[69] as approximately 50 hp-hours per tonne of municipal refuse or crop refuse ground. This is equivalent to about 136·4 MJ(th) per tonne of ground material. In his design for the bioconversion of wood chips to methane, Fraser[42] estimated energy requirements of 1230·5 MJ(th) per tonne of dry woody material. Taking the former figure as representative of leafy crop feed, the energy requirements for grinding will be

$$4380 \times 136\cdot4 = 0\cdot6 \times 10^6 \text{ MJ(th) per day}$$

(b) Compression and heat. Compression and heat account for the remaining part of the 7·5% of product methane used for shredding, compression and heat.[29] On a production basis of 50×10^6 cubic feet of methane per day at 10^3 Btu per cubic foot of methane, the energy requirements for compression and heat are estimated as

$$= 0\cdot32 \times 10^6 \text{ MJ(th) per day}$$

(c) Utilities. Sitton and Gaddy[29] give the annual cost of power, possibly for lighting and other utilities, as \$ 0·26 million per year at 1974 price levels. At 1967 price levels this can be put as \$ 0·133 million per year. Peters and Timmerhaus[76] give the cost of self-generated electricity as approximately \$ 0·015/kWh(e) at 1967 price levels. Thus from these figures*, the energy requirement of this input can be put as

$$\left(\frac{0\cdot133 \times 10^6}{330}\right) \times \left(\frac{3\cdot6 \times 3\cdot3}{0\cdot015}\right) = 0\cdot32 \times 10^6 \text{ MJ(th) per day}$$

* 1 kWh(e) = 3·6 MJ(e).

Land and water
These will be given zero energy factor to conform with the convention in ref. 72

Chemicals
Monoethanolamine solution (MEA) and glycol, probably triethylene glycol (TEG), are used respectively in the gas cleaning and drying units. The carbon dioxide in the gaseous product of bioconversion is scrubbed off with MEA solution while the TEG solution is used to dry the product methane. The costs of these are assumed to be included in those of the strippers and dryers.

Collection of crop material
This will involve the labour costs and energy inputs of collecting the crop materials, and since labour is given zero energy, by convention, the costs incurred will be treated in the final stage of cost estimation. The energy requirements for waste crops collection will certainly be less than those for specially grown crops because the latter will involve factors such as support energy and field operations energy.

(a) Crop wastes. A process plant based on crop wastes as feedstock should be located about 30 kilometres from farmlands where these wastes are to be collected. Farmers can collect these wastes and be paid for them. About 0·30 pence can be given for each kilogram of wastes giving rise to a waste cost of £ 3·0 per tonne of wastes. Thus for 4380 tonnes of wastes per day, the estimated cost of collection should be £ 13 140·0 per day. Transportation energy requirements will be similar to those of specially grown crops for the liquefaction process, i.e. the fuel energy for truck energy requirements

$$= 12 \text{ MJ(th) per vehicle-kilometre}$$

and for the truck body material contribution, the energy requirement

$$= 245 \text{ MJ(th) per truck-day}$$

Therefore for collection of 4380 tonnes of wastes per day, at a truck speed of 48 kilometres per hour (Appendix B5), transport energy requirements are

trip*: $48 \times 12 \times 30 = 0·02 \times 10^6$ MJ(th) per day
material: $245 \times 48 = 0·012 \times 10^6$ MJ(th) per day
Total: $= 0·032 \times 10^6$ MJ(th) per day

* 48 trucks will be required for the daily job.

Labour costs of truck drivers will be about

$$12 \times 1 \cdot 3 \times 48 = £\,750 \text{ per day}$$

where the average wage of an agricultural worker is £ 1·30 per hour.[77]

(b) Grown crops. Transportation energy requirements will be the same, i.e.

$$0 \cdot 032 \times 10^6 \text{ MJ(th) per day}$$

Transportation labour cost will also be about

$$£\,750 \text{ per day}$$

Support energy for the types of crops used, e.g. grass, can be assumed to be 3850 MJ per tonne-year[37] (including energy for tractor fuels and tractor material). Thus for a daily need of 4380 tonnes of crop material, the support energy required will be

$$4380 \times \frac{3850}{330} = 0 \cdot 051 \times 10^6 \text{ MJ(th) per day}$$

For field operations labour, assumptions similar to those in Appendix B5 will be made, i.e. about 260 days will be worked on the fields due to the winter period consideration. About 57 820 hectares of land will be required to give a daily crop requirement of 4380 tonnes. About 810 tractors will be required for this operation giving rise to labour costs of £ 3646 per day, for every 6·7 years. In this case additional capital investments will be required for the trucks and tractors. For 48 trucks at an estimated cost of £ 4000 per truck the additional investment will be £ 0·192 million. For 810 tractors at an average cost of £ 5500 each the additional investment is about £ 4·46 million. The GER of the bioconversion process is thus

for crop wastes $= 71 \cdot 697 \times 10^6$ MJ(th) per day
for grown crops $= 71 \cdot 75 \times 10^6$ MJ(th) per day

Outputs
No indication is made in the original design of the ratio of carbon dioxide to methane formed, but Hungate[43] indicates that the process of anaerobic digestion occurs at ordinary temperatures and pressures with a conversion efficiency as high as 94·0% on a thermal basis. Though fairly high, this efficiency of methane production will be assumed and 94% of the total GER of the process is attibuted to methane production. Therefore

$$\text{GER of methane} = 71 \cdot 70 \times 10^6 \times 0 \cdot 94$$
$$= 67 \cdot 4 \times 10^6 \text{ MJ(th) per day}$$

The CV of methane is about $1 \cdot 054$ MJ(th) per scf of gas.[30] Therefore

$$\text{CV of } 50 \times 10^6 \text{ scf of methane per day} = 52 \cdot 7 \times 10^6 \text{ MJ(th) per day}$$

The net energy requirement of the process

$$= \text{GER} - \text{CV of product}$$
$$= (67 \cdot 4 - 52 \cdot 7) \times 10^6$$
$$= 14 \cdot 7 \times 10^6 \text{ MJ(th) per daily scf}$$

and X_f value of process

$$= \frac{\text{NER of product}}{\text{CV of product}}$$
$$= \frac{14 \cdot 7 \times 10^6}{52 \cdot 7 \times 10^6} = 0 \cdot 28$$

The thermal efficiency of the process, with respect to methane,

$$= \frac{\text{output energy}}{\text{input energy}}$$
$$= \frac{52 \cdot 7 \times 10^6}{67 \cdot 4 \times 10^6}$$
$$= 78 \cdot 0\%$$

Energy Requirements in Bioconversion Processes: Woody Material Digestion

Inputs

Fraser[42] gives total plant investments as functions of number of pre-treatment–digestion trains, i.e. the number of digesters, steeping tanks and pH control units, that are required in the process; a train number of 16 is considered optimum by him. A plant with this level of equipment requirements would produce about 92.4×10^6 scf of methane per day, have an investment of \$(1974)250 million and would process 9430 tonnes of woody material per day.

At 1978 price levels the investments will amount to \$(1968) 141 million. Following the procedure in Appendix B8, the energy requirement of plant equipment

$$= \frac{1.5 \times 141 \times 10^6 \times 220}{2.4 \times 20 \times 330}$$
$$= 2.94 \times 10^6 \text{ MJ per day}$$

Crop material feed
Fraser[42] used wood species such as hybrid poplar, cotton woods and sycamore. These fall within the range of plants with a fuel value of about 15.40×10^6 J per kilogram of material.[29] Thus for 9430 tonnes per day, the energy input to the process from wood material

$$= 148.0 \times 10^6 \text{ MJ per day}$$

Power
(a) For grinding. About 14 hp-days per tonne of oven-dry material are required for grinding and shredding wood materials.[42] This is equivalent to

about 1.00 MJ per kilogram of wood, and hence power for grinding 9430 tonnes

$$= 9.60 \times 10^6 \text{ MJ per day}$$

(b) Compression and other utilities. No indication of these requirements is given in the original design but, following the estimate of Sitton and Gaddy[29] in Appendix B8, one could put these at roughly

$$6.4 \times 10^6 \text{ MJ(th) per day}$$

Hot water or steam
Hot water or steam at 190 °C (373 °F) and 180 psia is required to solubilise the wood chips in the steeping tank.[42] From steam tables, saturated steam has a heat content of 1187 Btu per pound or 2802 MJ per tonne. From Perry[48] saturated steam at 190 °C (373 °F), 180 psia, has the following properties: heat content of vapour is 1187 Btu per pound or 2802 MJ(th) per tonne; heat content of liquid water in equilibrium with vapour is 346 Btu per pound or 817 MJ(th) per tonne. Therefore the heat required to raise 1 tonne of steam from water at 21 °C (70 °F), standard conditions, will be 2802 MJ(th). Fraser[42] indicates that the slurry composition in the steeping tank should be about 15% solids by weight. Thus for solubilisation of 9430 tonnes of wood each day about 53 438 tonnes minimum of water are required for steam purposes, or raising to boiling point. Therefore

energy requirement for steeping hot water $\simeq 817 \times 53\,438$

$$= 43.7 \times 10^6 \text{ MJ(th) per day}$$

Land and water
As in Appendix B8 these will be given zero energy value.

Chemicals
The cooled slurry (15% solids composition) is passed to the pH control unit where fixed nitrogen or phosphate is put to raise the pH value slightly above neutral. About 1 tonne of fixed nitrogen per 100 tonnes of wood chips is required[42] giving a daily requirement of chemicals of 94.3 tonnes.

From ECN,[49] the market prices of ammonium persulphate, ammonium sulphate and phosphoric acid, food grade, are respectively 33 US cents per pound, 3.72 cents per pound and 17 cents per pound. An average value of 20 cents will be chosen for the purpose of this work. This is about £ 236 per tonne of chemicals at 1978 price levels or, assuming an average inflation factor of 10% per year, £ 91.0 per tonne at 1968 price levels.

From Casper *et al.*,[72] the energy requirements of chemicals are about 702 MJ per £ (1968). Therefore

$$\begin{aligned}
\text{energy input to process by way of} \\
\text{chemical addition} = 702 \times 91 \cdot 0 \times 94 \cdot 3 \\
= 6 \cdot 02 \times 10^6 \text{ MJ(th) per day}
\end{aligned}$$

Collection of crop material

The estimate of energy requirements under this subsection will be quite rough. In the original design, Fraser[42] indicates that about 175 million acres (0·71 million square kilometres) of US land will be needed to yield deciduous woody material for conversion to methane. At a growth rate of about 2 kilograms per metre^2-year of plant materials and a conversion potential of 4·5 scf of methane gas per pound of dry woody material (0·281 m^3/kg), about 14×10^{12} scf of SNG per year is expected. For the US this is quite feasible from the viewpoint of land availability, but for countries such as the UK this is going to be a limiting factor. Nonetheless, the theoretical energy requirements will be estimated.

A 2 kilogram per metre^2-year plant growth rate will need about 154 490 hectares (1 hectare $\simeq$ 10 143 square metres) to make available 9430 tonnes of plant material. Transportation mode will be same as in the case of leafy crop material, i.e. with trucks of fuel consumption of 12 MJ(th) per vehicle-kilometre and material of construction energy equivalent of 245 MJ(th) per truck-day. A collection of 9430 tonnes of wood per day will require about 104 theoretical trucks, i.e. the energy requirements in vehicle use will be

for trips $104 \times 12 \times 30$ $= 0 \cdot 04 \times 10^6$ MJ per day
for vehicle material 245×104 $= 0 \cdot 03 \times 10^6$ MJ per day
Total $0 \cdot 07 \times 10^6$ MJ per day

From Sitton and Gaddy[29] it will be seen that the estimated solar energy conversion percentages of sycamore, hybrid poplar and some grass species are quite similar, i.e. about 0·24–0·60. With this in mind one can conveniently put the support energy required for growth and collection of plant material as approximately that of grass, i.e. about 4000 MJ per tonne-year. Hence for a daily need of 9430 tonnes of material, the support energy requirements are

$$9430 \times \frac{4000}{330}$$

$$= 0 \cdot 114 \times 10^6 \text{ MJ(th) per day}$$

Labour costs and capital costs in the collection and field operations will be similar to those presented in Appendix B5. With a land requirement of 154 490 hectares, about 2164 theoretical tractors will be required over a period of 6 years, each tractor working 900 hours per year (of 260 days). A labour cost of £9740 per day is required for tractor drivers and £1622 per day for truck drivers. At an average cost of £4000 per truck and £5500 per tractor capital costs (initial vehicle and tractor investments)

$$\simeq (104 \times 4000) + (2164 \times 5500)$$
$$= £(1978)\ 12 \cdot 32\ \text{million}$$

The GER of the process is thus

$$10^6(2 \cdot 94 + 148 + 16 + 43 \cdot 7 + 6 \cdot 02 + 0 \cdot 26)$$
$$= 217 \cdot 0 \times 10^6\ \text{MJ(th) per day}$$

Outputs

At an average rate of 4·5 scf of methane per pound of woody material, the expected output of methane is 92·4 million scf per day. But in the original design, Fraser[42] puts the actual output of methane at 77 million scf and 77 million scf of carbon dioxide, based on a fifty–fifty ratio of methane–carbon dioxide production. However, in a sensitivity analysis the methane–carbon dioxide ratio is put at sixty–forty. Thus a yield of 92·4 million scf of methane per day is expected. Therefore

$$\text{GER of process attributable to methane} = 217 \times 10^6 \times 0 \cdot 6$$
$$= 130 \cdot 12 \times 10^6\ \text{MJ(th) per day}$$

The CV of methane is about 1·054 MJ per scf.[30] Therefore

$$\text{heat value of product} = 1 \cdot 054 \times 92 \cdot 4 \times 10^6$$
$$= 97 \cdot 4 \times 10^6\ \text{MJ per day}$$

The NER of methane product is given by

$$(\text{GER} - \text{heat value})$$
$$= (130 \cdot 12 - 97 \cdot 4) \times 10^6$$
$$= 32 \cdot 72 \times 10^6\ \text{MJ per day}$$

$$\text{and the } X_f \text{ value} = \frac{\text{NER}}{\text{heat value}}$$
$$= \frac{32 \cdot 72 \times 10^6}{97 \cdot 4 \times 10^6}$$
$$= 0 \cdot 336$$

Evaluation of (f) Values for Various Coal Processes

From the cost equation for coal processes, energy costs are attributed to

$$C' = (1 + X_f) \, P'_f \, (f)$$

where C' is the energy cost element of total cost production (£ per tonne of product oil), X_f is the ratio of NER of process to calorific value of product oil and P'_f is the price of coal (£ per tonne of coal). The (f) factor is thus defined as the calorific value of the product oil divided by the calorific value of the coal feedstock. In units this is equal to

$$\frac{MJ}{\text{tonnes of product oil}} \cdot \frac{MJ}{\text{tonnes of coal}}$$

$$= \frac{\text{tonnes of coal}}{\text{tonnes of product oil}}$$

Taking the H-coal process (case 1) as an example, input coal has a calorific value of 25 530 MJ per tonne while the product oil has value of 42 700 MJ per tonne. The (f) value of the process is given by

$$\frac{42\ 700}{25\ 530} = 1.68$$

The (f) values of other coal processes may be calculated in a similar way, and these are shown in Table 47.

APPENDIX C2

Evaluation of Capital Costs, $X_c P_c$, for Various Coal Processes

By definition, X_c is the quantity of capital required to produce 1 tonne of syncrude or product. This is equivalent to the capital investment of the process. P_c is the cost of capital which can be translated as the annual charges rate in percentages. However, in most coal processes more than one product is formed. Therefore, depending on the major products required, the total investments may not be fully apportioned to one product only. A factor D_c is introduced to allow for this distribution of capital; it is equivalent to the fraction of the product under consideration of the total major products of the process. In the case of the NCB hydrogenative extraction process, where the three major products are considered together, D_c is taken as unity. Also in the Fischer–Tropsch process of motor spirit product, D_c is taken as unity. The essence of D_c is to distribute capital to products in terms of energy value.

Taking the H-coal process as an example, Goen et al.[70] estimate the cost of the process plant producing about 14 452 tonnes of syncrude per day as (1974) 533 million. At 1978 price levels this is equivalent to

$$\frac{533 \times 10^6 \times (1\cdot1)^4}{1\cdot95}$$

$$= £(1978) \ 400 \text{ million}$$

where £(1978)1·00 is taken as $(1978)1·95 and $(1\cdot1)^4$ is the factor of inflation. The value of $X_c P_c$ of the process can then be defined as

$$X_c P_c = R'_c \ (\text{CI}) \ D_c$$

where R'_c is the project annual interest charges rate, given as a fraction, and CI is the corrected capital investment at 1978 price levels. From DCF calculations with project lifetime taken as 20 years, 10% capital charges rate will be adequate. Another 5% will be charged for maintenance and utilities services. Therefore R'_c is equal to 0·15. Hence $X_c P_c$ of the process

$$= 0·15 \ (400 \times 10^6) \times 0·95$$
$$= £\ 57 \text{ million}$$

But the daily product yield is 14 452 tonnes. This is equivalent to 14 452 × 365 × 0·9 tonnes per year of 90% working period. Therefore

$$X_c P_c \text{ per tonne of product syncrude} = \frac{57 \times 10^6}{14 \ 452 \times 365 \times 0·9}$$
$$= £\ 12·00$$

This means that to produce 1 tonne of syncrude from coal, using the direct hydrogenation process, will require £ 12·00 at 1978 price levels. In a similar manner the values of $X_c P_c$ per tonne of product are calculated for some other processes. These are tabulated in Table 49.

Evaluation of Labour Costs, X_1P_1, for Various Coal Processes

In Section 9.4 the value of P_1, the price of labour (manual average) in the UK, was calculated as £1·85 per man-hour worked. A mean value of labour input to coal processing establishments, such as the NCB Research Centre at Cheltenham, was given as 1·04 million man-hours per year. The factor D_1 was also introduced to account for labour distribution to the various products on an energy value basis. D_1, on this ground, is equivalent to D_c. In this appendix the values of D_1 are given and these are used to calculate various labour costs, X_1P_1, of the products. The values of X_1P_1 are given in Table 51.

H-coal (Case 1),	D_1 value = 0·95
H-coal (Case 2),	D_1 value = 0·96
COED (Case 1),	D_1 value = 0·40
COED (Case 2),	D_1 value = 0·35
CSF(Case 1),	D_1 value = 0·97
CSF (Case 2),	D_1 value = 0·97
Fischer – Tropsch (motor spirit),	D_1 value = 1·00
Fischer – Tropsch (methanol),	D_1 value = 0·66
NCB, liquefaction,	D_1 value = 1·00
NCB, supercritical extraction,	D_1 value = 1·00

APPENDIX C4

Estimates of Costs of Producing Pipeline Gas from Coal by Pyrolysis

CAPITAL COSTS

Shearer's[17] design is based on pyrolysis of coal coupled with the gasification of the resulting char. In Part I it was found that 57·85% of the total GER of the process is attributable to the production of pipeline gas. In calculations similar to those presented in Appendix C2, it can be shown that the annual capital charges of plant producing such quantities of gas are approximately given by

$$X_c P_c = \frac{482 \cdot 9 \times 10^6 \times (1 \cdot 1)^4 \times 0 \cdot 5785 \times 0 \cdot 15}{1 \cdot 95 \times 249 \cdot 6 \times 10^3 \times 365 \times 0 \cdot 9}$$
$$= £ 0 \cdot 38 \text{ per 1000 scf of gas}$$

where $482 \cdot 9 \times 10^6$ is the $(1974) investment in plant and $(1978)1·95 = £(1978)1·0, 7·08 × 10^6 cubic metres of gas are equivalent to $249 \cdot 6 \times 10^6$ cubic feet, 0·5785 is the fraction of total plant cost apportionable to gaseous products in terms of energy value, and 0·15 is the annual capital charges rate. It is also assumed that the plant is on-stream 90% of the time in a year and the annual inflation rate is 10%.

LABOUR COSTS

The labour cost factor, $X_1 P_1$, can also be obtained in calculations similar to those in Appendix C3 and Section 9.4. In this case the $X_1 P_1$ value is approximately

£ 0·014 per 1000 scf of gas

230

ENERGY COSTS

The fuel energy cost factor will also be calculated in the usual manner, viz:

$$\text{energy cost} = X_f P_f'(f)$$

where $X_f = \dfrac{\text{NER of product}}{\text{heat value of product}}$
      $= 0 \cdot 812$ (calculated in Part I)
      $X_f' = 1 \cdot 812$
      $P_f' = $ price of coal (£ per tonne)
      $= £ (1978) \, 22 \cdot 60$ per tonne
and $(f) = $ the ratio of the calorific value of product per 1000 scf of pipeline gas to coal used.

In Shearer's design, 29 410 tonnes equivalent of coal per day are required to yield 7·08 million cubic metres or 249·6 million cubic feet of gas per day. This quantity of gas accounted for 57·85% of total products in terms of energy yield. The calorific value of coal was 26 040 MJ per tonne. Therefore the (f) value of gas product is given by

$$f = \frac{1 \cdot 05 \times 10^6 \times 1000}{26\,040 \times 10^6}$$

$$= 0 \cdot 04$$

Therefore

$$\text{energy cost, } X_f P_f'(f) = 0 \cdot 04 \times 0 \cdot 812 \times 22 \cdot 60$$
$$= £ \, 1 \cdot 65 \text{ per 1000 scf of gas}$$

The total costs of producing pipeline gas by this process are given in Table 61.

Evaluation of Capital Costs, $X_c P_c$, for Plants Processing Tar Sands into Synthetic Crude

Swabb[6] states that the initial investment at the GCOS was about \$253 million, at 1967 price levels, for a plant of capacity of about 2·2 million tonnes of synthetic crude per annum. Hemming[50] gives the investment as \$(1964)171 million for a plant with a capacity of 2·3 million tonnes per annum. At 1978 price levels these figures are quite similar. However, using the figure \$(1974)171 million for a throughput of 2·3 million annual tonnes the estimated capital charges per annum are calculated by

$$X_c P_c = R_c' (CI)$$

The value R_c' will be taken as 15% or 0·15. Therefore

$$X_c P_c \text{ of plant} = \frac{171 \times 10^6 \times (1·1)^4 \times 0·15}{1·95}$$
$$= £(1978) \ 51·27 \text{ million}$$

But the product yield from plant is 2·3 million tonnes. Hence

$$X_c P_c \text{ per tonne of product} = \frac{51·27 \times 10^6}{2·3 \times 10^6}$$

$$= £22·30 \text{ per tonne of product}$$

Refuse Collection Costs

Refuse collection is a very vital aspect of the process of getting synthetic oil from municipal refuse. It can determine to a great extent the actual costs of processing the rubbish. Many municipalities have various patterns of collecting their refuse. In the UK most county councils collect their refuse in trucks specially designed for this purpose. Capacities of trucks vary from about 20 cubic yards to about 60 cubic yards. The Greater Manchester County Council employs solely this method of collection. The Greater London area employs the same method of collection though plans are underway to introduce rail haulage for refuse.[67] The Greater London area is constructing four railway lines to help the council dispose of refuse at a faster, and supposedly cheaper rate.

However, the mode of collection here is assumed to be by trucks. The local population can help the various councils by engaging in a segregated pattern of disposal. In actual fact, some councils are thinking of introducing bonus schemes to tenants who are willing to engage in a selective pattern of disposal. In this way it is intended that cellulosic materials such as paper and vegetable matter are disposed of separately from packaging materials such as plastics and solid, non-combustible materials such as bottles and tins. This will enable local councils to dispose of combustible materials easily, without the problem of sorting. If this plan is accepted then refuse collection can be done more easily, and at a cheaper rate, than it is done today. Because of the non-availability of published reliable information on this aspect of waste treatment the works of Bidwell and Mason[34] on the Greater London area will be used to ascertain labour and other costs of refuse collection. For ease of calculation this refuse collection cost will be treated as labour costs together with plant process operational staff labour costs. According to Mason and Bidwell, the cost of transporting 1 tonne of refuse over 1 km in the Greater London area, as of 1975, was 4 pence. Suppose over the years an inflationary factor of 10% is introduced and the cost, as of 1978, is taken as 6 pence per tonne-kilometre.

An average distance of 30 km is assumed to be the one-way journey of every tonne of refuse collected. Then for every tonne of refuse a collection cost of £1·80 is required by the GLC. This appears to be quite representative in the UK. The Greater Manchester County Council quotes figures which range from £1·60 to £1·80 per tonne of refuse collected. If this figure of £1·80 is assumed then the daily costs of refuse collection for a 36 tonnes per day plant will be equivalent to

$$6·0 \times 48 \times 30 = £1·80 \times 48$$
$$= £86·40 \text{ per day}$$

The figure 48 originates because, as in the Holden Town case, the municipal refuse is assumed to be 75% combustible materials and 25% non-combustibles. In actual fact, Kaufman and Weiss[25] give the Holden Town refuse composition as above. Thus to get 36 tonnes of combustible refuse, about 48 tonnes of total refuse have to be collected.

Cost Estimates for Liquefaction Plants with Capacities Greater than 36 tonnes per day

It is necessary to look at plants with capacities greater than 36 tonnes per day because to be of commercial importance a plant should be treating about 2000 tonnes of refuse per day because of the rate of refuse disposal in large populations. Essentially two cost elements will change with increase in capacity.

First, the capital costs per tonne of product oil will change because of the six-tenths rule to plant costs estimation. As an example, a 36 tonnes per day plant will cost $ 1·49 million to build while a 100 tonnes per day plant will cost $ 2·74 million to build. In terms of product yield the 36 tonnes plant will produce 13 tonnes of oil each day while the 100 tonnes plant will produce 36 tonnes of oil. Therefore the capital charges per tonne of product oil will be about $ 59·00 for a 36 tonnes plant while for a 100 tonnes plant the charges per tonne of product oil will be $ 39·00. This represents quite a change in capital cost factor.

Secondly the direct labour input to the processing plant can be assumed to remain fairly constant. But an increase in capacity will mean an increase in product yield and a decrease in labour costs per tonne of product oil. From Appendix C6 it will be seen that transportation costs of refuse will remain the same at £ 1·80 per tonne of refuse for a 30 km journey. Also, since input refuse constitutes about 87% of the total energy input to the process the energy cost factor will tend to remain fairly linear with plant capacity. The various cost factors have been estimated for the various plant capacities and these are shown in Table 59 and their contributions to the total costs of producing 1 tonne of oil from refuse are given in Table 60. Note that the least economic capacity turns out to be about 500 tonnes of combustible refuse per day (see Fig. 27).

APPENDIX C8

Cost Estimates for Pyrolysis Plants with Capacities Greater than 200 tonnes per day

As in the case of the refuse liquefaction plant it is necessary to estimate costs for pyrolysis plants, based on the Garret process, with capacities greater than 200 tonnes per day. Capital charges are treated in the same way with a charges rate of 17% annually and the use of the six-tenths rule for scaling up costs with capacities.

Labour costs will also vary with capacities. In this case the variation is an inverse proportion type of variation. One significant point in this process is that the energy factor, X_f', is quite high, about 0·70, and this tends to make the cost of energy quite comparable to other costs. Table 48 is calculated and presented based on the assumptions made above. It will be seen that quite a gain is made in production costs when capacities move from 200 to about 2000 tonnes per day.

Comparative Analyses and Projections of Fuel Costs in the UK

NATURE-MADE PETROLEUM

In 1976 the price of nature-made petroleum was \$ 11·00 per barrel. In 1978, after a temporary price freeze by the OPEC nations, the price of oil was about \$ 14·00 per barrel. Now many producing countries hope to increase oil prices by about 6 – 15%. Based on the 1978 price of \$ 14·00 per barrel, the mean market price of oil in the UK could be put at 11·60 pence per therm. Suppose between 1978 and 2000 A.D. there is a mean annual inflation rate of 6%. One can then put petroleum prices in 2000 A.D. at 39 pence per therm or £ 24 (\$ 48) per barrel. To make for flexibility, the expected prices of petroleum in 2000 A.D. are given in Table 84 as functions of inflation rates. These projections are quite close to those of Barnea.[78]

GAS PRODUCTION

Cellulose digestion

The cost of producing 1000 scf of methane from cellulose by anaerobic

TABLE 84

Inflation rates per annum (%)	Estimated petroleum price in 2000 A.D. (£ per barrel)
5	19·50
6	24·00
8	35·00
10	52·00

TABLE 85

Annual inflation rate (%)	Estimated price of gas from cellulose in 2000 A.D. (pence per therm)[a]
5	57 (£ 5·70 per 1000 scf)
6	71
8	106
10	155

[a] Prices are put at 1·75 times cost of production.

digestion is approximately £ 1·20 (Table 75). This is based on 1979 price levels and is equivalent to approximately 12 pence per therm. Suppose inflation rates are applied to both capital and labour cost elements of the conversion process, then the data shown in Table 85 are obtainable.

Coal – char gasification

Synthetic natural gas can be obtained from coal presently at an estimated cost of 17 pence per therm. Applying the annual inflation rates as above, the prices of coal-based SNG given in Table 86 can be obtained.

TABLE 86

Annual inflation rate[a]	Price of gas from coal in 2000 A.D. (pence per therm)[b]
5% on capital equipment, 2% on price of coal	52
6% on capital equipment, 2% on price of coal	57
8% on capital equipment, 2% on price of coal	67

[a] Coal is given 2% annual inflation, different from capital costs of equipment, because coal as a commodity is not a final product with respect to synthetics.
[b] Prices are put at 1·75 times cost of production.

SYNTHETIC OIL PRODUCTION

From coal

A typical liquefaction process will yield synthetic oil from coal at a production cost of £ 8·30 per barrel (Section 13.1). In the same manner of

TABLE 87

Inflation rate per annum	Price of coal-derived oil in 2000 A.D. (£ per barrel)
5% on capital and labour costs, 2% on coal prices	19·60
6% on capital and labour costs, 2% on coal prices	21·70
8% on capital and labour costs, 2% on coal prices	28·00

inflation rate application as that shown above, one can estimate the various prices of coal-derived oils in 2000 A.D. as shown in Table 87.

From cellulose
From estimated costs of producing oils from cellulose (Chapter 11) (refuse and grown plants considered together), it is possible to derive synthetic crude from cellulosic materials at a mean cost of £ 24·00 per tonne of oil or £ 3·40 per barrel. In the case of cellulose-based processes the usual practice (throughout this work) is to apply inflation rates to only capital and labour cost factors. The labour cost here involves both collection and operating staff costs. Thus applying varying inflationary factors the values in Table 88 can be obtained.

TABLE 88

Inflation rate per annum (%)	Price of cellulose-derived oil in 2000 A.D. (£ per barrel)
5	9·50
6	11·60
8	17·10

DISCUSSION

As of 1979, nature-made petroleum is cheaper than any anticipated coal-derived synthetic crude, with typical prices of about £ 7·0 to £ 8·50 per barrel. But with time the reverse may be the case. The tendency for petroleum prices to rise sporadically may make coal-derived oils more and more competitive. Cellulose-derived oils will always be as competitive as nature-made petroleum. In the case of gas, as long as natural gas exists in abundance there is a grave doubt whether SNG from coal will be cheaper. However, as nature-made gas decreases in reserves, substituted natural gas will become very competitive in price. Note that an average annual rate of inflation of 2% has been given to coal prices. If, however, the rate of inflation surpasses this then coal-derived gases and oils may be even less competitive with time.

References

1. P. F. Chapman, 'The Role of Energy in Waste Management Policies'. A report prepared by the Open University Research Group for the WMAC Economics Group, 1976.
2. D. W. Boltfield, 'The Energy Audit in Practice'. Symposium organised by the Institution of Chemical Engineers on Practical Energy Saving, Manchester, England, 15 March 1978.
3. P. F. Chapman, 'Principles of Energy Analysis'. A reprint from *Aspects of Energy Conversion*, Blair, Jones and Van Horn (eds.), Pergamon Press, Oxford, 1976.
4. 'Energy Outlook 1977-1990'. A publication of Exxon Co., USA, January 1977.
5. 'Scenario 0: Availability and Costs of Energy in the UK'. A paper prepared for the Energy Technology Support Unit, R & D Purposes, Abingdon, 2 December 1977.
6. L. E. Swabb, Jr, 'Synthetic Fuel Activities in the Western Hemisphere'. Annual Convention of the German Society for Mineral Oil Science and Coal Chemistry, Hamburg, 30 September-3 October 1974.
7. C. W. Bowman and M. A. Carrigy, 'World-wide Oil Sand Reserves'. Conference on the Future Supply of Nature-made Petroleum and Gas, Laxenburg, Austria, 5-16 July 1976, pp. 732-44.
8. W. L. Oliver, 'An Overview of Canada's First Commercial Surface Mining Extraction Plant'. idem, pp. 754-62.
9. L. C. Marchant, 'Activity on the Production of Energy from the Tar Sands of the United States'. idem, pp. 763-78.
10. 'Plastics Waste Disposal: Some Important Considerations'. A report on plastic disposability by The Plastics Institute, 11 Hobart Place, London, HMSO, April 1971.
11. D. Hemming, 'Energy Requirements for the Conversion of Coal into Liquid Fuels'. Open University Research Report No. ERG 017, May 1977.
12. D. Hemming, 'Energy Requirements for the Production of Synthetic Crude Oil

242   *Production and utilisation of synthetic fuels—An energy economics study*

from Athabasca Tar Sands'. Open University Research Report No. ERG 015, November 1976.
13. D. Hemming, 'Energy Requirements for the Production of Synthetic Crude Oil from Colorado Oil Shales'. Open University Research Report No. ERG 012, May 1976.
14. *Coal Processing Technology*, Vol. 3, edited by Chemical Engineering Progress and published by American Institute of Chemical Engineers, New York, 1977, p. 142.
15. N. P. Cochran, 'Oil and Gas from Coal'. *Scientific American*, **234**, May 1976, p. 24.
16. Consolidation Coal Company, 'Project Gasoline', Vol. 1. Summary Report, August 1969, available from US NTIS, reference number PB-234 125/3GA.
17. H. A. Shearer, 'The COED Process Plus Char Gasification'. *Chemical Engineering Progress*, **69**, No. 3, March 1973, p. 43.
18. R. T. Eddinger, 'A Synthetic Fuels Industry Based on Rocky Mountain Coals'. *Quarterly Report of the Colorado School of Mines*, **65**, No. 5, October 1970, p. 183.
19. F. L. Chan *et al.*, 'A SASOL-type Process for Gasoline, Methanol, SNG and Low-Btu Gas from Coal'. Available from US NTIS, reference number PB-237 670/5GA.
20. G. O. Davies *et al.*, 'A Route to Hydrocarbon Liquids by the Hydrogenation of Solvent Extracts from Coal'. Symposium on Gasification and Liquefaction of Coal, Working Meeting Number II, Dusseldorf, 12–16 January 1976.
21. R. R. Maddocks and J. Gibson, 'Coal Processing: Supercritical Extraction of Coal'. A report prepared for the NCB by the Coal Utilisation and Research Unit, London, 1975.
22. D. Hemming, 'The Relative Costs of Producing Syncrude from Oil Shales of Various Grades Under Varying Conditions'. Open University Research Report No. ERG 022, January 1978.
23. D. Hemming, 'Labour and Capital Requirements of Syncrude Produced from Oil Shales in an Aboveground Retorting Facility'. Open University Research Report, November 1977.
24. P. F. Chapman, 'The Energy Costs of Producing Copper and Aluminium from Primary Sources'. Open University Research Report No. ERG 001, December 1973.
25. J. Kaufman and A. H. Weiss, 'Solid Waste Conversion: Cellulose Liquefaction'. A report prepared for the US National Environmental Research Centre by Worcester Polytechnic Institute, 1975. Distributed by NTIS, Department of Commerce, Springfield, USA.
26. D. E. McCloud *et al.*, 'A New Look at Energy Sources'. American Society of Agronomy Special Publication No. 22, Madison, Wisconsin, 1974.
27. C. C. Kemp and G. C. Szego, 'The Energy Plantation'. Proc. AICLE Symposium on Solar Energy Utilisation, 20 March 1975.
28. 'Energy Alternatives: A Comparative Analysis'. A Report on Organic Wastes Resources Systems, No. FEA/D-75/551, for the US Federal Energy Administration, 1975. Reproduced by NTIS, Department of Commerce, Springfield, USA.
29. O. C. Sitton and J. L. Gaddy, 'Solar Energy Collection by Bioconversion'. 11th

Proceedings of the IECEC, University of Missouri, Rolla, Missouri, USA, 1975, p. 91.

30. S. S. Penner and L. Icerman, *Energy. Vol. I. Demands, Resources, Impact, Technology and Policy.* Addison-Wesley Publishing Co. Inc., Mass., 1974, pp. 1–25.

31. Alich and Inman, 'Effective Utilisation of Solar Energy to Produce Clean Fuels'. US National Science Foundation, RANN, Final Report, 1974.

32. L. L. Anderson, 'Energy Potential from Organic Wastes: A Review of the Quantities and Sources'. US Department of the Interior, Bureau of Mines Information Circular No. 8549, Washington, DC, 1972.

33. I. Schomburgh, 'The Development of Conversion of Refuse to Energy Innovation in Britain'. National Research Development Corporation, London.

34. R. Bidwell and S. Mason, 'Fuel from London's Refuse: An Examination of Economic Viability'. Environmental Resources Ltd, London, 1974.

35. J. P. Cooper, 'Photosynthesis as an Industrial Energy Source?'. Welsh Plant Breeding Station, Aberystwyth, published in *Energy and the Environment*, J. Walker (ed.).

36. D. E. Earl, *Forest Energy and Economic Development.* Clarendon Press, Oxford, 1975, pp. 43–75.

37. G. Leach, *Energy and Food Production.* IPC Science and Technology Press Ltd, Surrey, England, 1976.

38. B. J. Flanagan, 'Pyrolysis of Domestic Refuse with Mineral Recovery'. Oxy Metal Industries, Birmingham, UK. Paper available in CRE, 1st International Conference and Technical Exhibition on Conversion of Refuse to Energy, Montreux, Switzerland, 3–5 November 1975, p. 220.

39. S. J. Levy, 'The Conversion of Municipal Solid Wastes to a Liquid Fuel by Pyrolysis'. Office of Solid Waste Management Programme, Environmental Protection Agency, Washington, DC. Paper available in CRE, 1st International Conference and Technical Exhibition on Conversion of Refuse to Energy, Montreux, Switzerland, 3–5 November 1975, p. 226.

40. C. R. Rice, 'Landfill Site Eyed as Methane Source'. *Chemical and Engineering News*, 28 August 1978, p. 25.

41. R. W. Graham *et al.*, 'A Preliminary Assessment of the Feasibility of Deriving Liquid and Gaseous Fuels from Grown and Waste Organics'. 11th Proceedings of the IECEC, University of Missouri, Rolla, Missouri, USA, 1975, p. 98.

42. M. D. Fraser, 'Solar SNG: Large-scale Production of SNG by Anaerobic Digestion of Specially Grown Plant Matter'. 11th Proceedings of the IECEC, University of Missouri, Rolla, Missouri, USA, 1975, p. 83.

43. R. E. Hungate, 'Potentials and Limitations of Microbial Methanogenesis'. *ASM News*, **40**, No. 11, 1974.

44. L. G. Massey, speaking at the Symposium on Coal Gasification, sponsored by the Division of Fuel Chemistry of the American Chemical Society, 165th Meeting.

45. D. Merrick, 'Advanced Coal Conversion Processes'. National Coal Board Research Establishment, Cheltenham, 1978.

46. *Coal Processing Technology*, Vol. 2, edited by Chemical Engineering Progress and published by American Institute of Chemical Engineers, New York, 1975.

47. S. S. Penner and L. Icerman, *Energy. Volume II. Non-Nuclear Technologies.* Addison-Wesley Publishing Co. Inc., Mass., 1975, p. 99.
48. *Digest of UK Energy Statistics, 1977.* Prepared by the Department of Energy and the Central Office of Information, London, UK, 1977.
49. *Industrial Research in Britain*, 1976 Edition. Published by Francis Hodgson Ltd, Guernsey.
50. D. Hemming, 'Energy Requirements for the Production of Synthetic Crude Oil from Athabasca Tar Sands'. Open University Research Report No. ERG 022, January 1978.
51. J. R. Donnel, 'Global Oil-Shale Resources and Costs'. Conference on the Future Supply of Nature-made Petroleum and Gas, Laxenburg, Austria, 5–16 July 1976, p. 843.
52. C. H. Prien, 'Current Oil Shale Technology: Summary from Guidebook to the Energy Resources of the Piceance Creek Basin, Colorado'. Rocky Mountain Association of Geologists, Denver, Colorado, 1974, p. 141.
53. *Energy Trends.* A Statistical Bulletin of the UK Department of Energy, June 1978, available at Central Office of Information, London.
54. *Canada Yearbook 1974.* An annual review of economic, social and political developments in Canada, published by the Ministry of Industry, Trade and Commerce, Ottawa, p. 315.
55. W. L. Oliver, 'Global Oil-Shale Resources and Costs'. Conference on the Future Supply of Nature-made Petroleum and Gas, Laxenburg, Austria, 5–16 July 1976, p. 745.
56. Amendments to the Agricultural Wages Act of 1948, operative 20 January 1978, HMSO, London, 1978.
57. W. J. Most and E. E. Wigg, 'Methanol and Methanol–Gasoline Blends as Automotive Fuels'. Meeting of the Combustion Institute, Central States Section, Columbus, Ohio, 5–6 April 1976.
58. *United Kingdom Energy Statistics 1977.* A publication of The Department of Energy, Economics and Statistical Division, HMSO, London, 1979.
59. *Energy Trends.* A statistical bulletin of the UK Department of Energy, Economics and Statistical Division, January 1979, available at Central Office of Information, London.
60. British Gas, Annual Report and Accounts 1976–77 of the British Gas Corporation, HMSO, London, 1977.
61. *Handbook of Electricity Supply Statistics*, 1978 Edition. Issued by the Intelligence Section, Secretary's Department, Electricity Council, London. Printed by Unwin Bros. Ltd, Surrey, England, pp. 26–31, 56–7, 80–93.
62. *Electrical Times Electricity Supply Handbook 1978.* A publication of the Electricity Council, London.
63. G. P. Hill, *Power Generation: Resources, Hazards, Technology and Costs*, MIT Press, Cambridge, Mass., 1977.
64. *The Guardian*, 20 February 1979, p. 20.
65. *Annual Abstract of Statistics 1977.* A publication of The Government Statistical Service, Central Statistics Office, HMSO, London, 1977.
66. L. Grainger, 'Coal into the Twentyfirst Century'. Robens Coal Science Lecture delivered at the Meeting of the British Coal Utilisation Research Association, London, 7 October 1974.

67. 'London Plans Rail Haulage for Refuse Disposal'. Article on Solid Waste Disposal, *Engineering Index*, April 1978, p. 476.
68. L. E. J. Robert, 'Energy and the Environment'. Symposium on Man and His Environment, held at University of Birmingham, September 1975, J. Walker (ed.), pp. 99–118.
69. 'Energy Resources'. A Second Level Course in Science, Open University Press, p. 24.
70. R. L. Goen, C. F. Clark and M. M. Moore, 'Synthetic Petroleum for Department of Defense Use'. Available as NTIS reference AD/A-005 403/1GA, 1974.
71. US Department of Commerce, Bureau of the Census, *Statistical Abstracts of the United States*. 96th Annual Edition, 1975, p. 418.
72. D. A. Casper *et al.*, 'Energy Analysis of the Report on the Census of Production, 1968'. Open University Research Report No. ERG 006, November 1975.
73. P. Harris, Energy Technology Support Unit, Harwell, UK, personal communication. Calculations made on data obtained from H. R. Englehardt, Jülich Report No. Jül-966-RG, June 1973.
74. J. O. Edewor, 'Energy Economics'. M.Sc. Dissertation, UMIST, 1977.
75. D. Pimental *et al.*, 'Workshop on Research Methodologies for Studies of Energy, Food, Man and Environment: Phases I and II'. Cornell University Centre for Environmental Quality Management, Ithaca, New York. Phase I, June 1974; Phase II, November 1974.
76. M. S. Peters and K. D. Timmerhaus, *Plant Design and Economics for Chemical Engineers*, 2nd Edition, McGraw-Hill Inc., 1968, p. 772.
77. J. H. Perry *et al.*, *Chemical Engineers Handbook*, 4th Edition, McGraw-Hill Inc., pp. 3–191.
78. J. Barnea. 'The Future Supply of Nature-made Petroleum and Gas'. Conference on the Future Supply of Nature-made Petroleum and Gas, Laxenburg, Austria, 5–16 July 1976, pp. 23–40.

Index